ÉLIPHAS LÉVI

The Key of the Mysteries

Translated from the French,
with an Introduction and Notes, by
ALEISTER CROWLEY

ISBN: 978-1-63923-455-4

Printed: June 2022

Cover Art By: Amit Paul

Published and Distributed By:
Lushena Books
607 Country Club Drive, Unit E
Bensenville, IL 60106
www.lushenabooksinc.com/books

ISBN: 978-1-63923-455-4

ÉLIPHAS LÉVI

The Key of the Mysteries

Translated from the French,
with an Introduction and Notes, by
ALEISTER CROWLEY

Contents

ILLUSTRATIONS

Introduction

THIS volume represents the high-water mark of the thought of Éliphas Lévi. He is no longer talking of things as if their sense was fixed and universal. He is beginning to see something of the contradiction inherent in the nature of things, or, at any rate, he constantly illustrates the fact that the planes are to be kept separate for practical purposes, although in the final analysis they turn out to be one. This, and the extraordinarily subtle and delicate irony of which Éliphas Lévi is one of the greatest masters that has ever lived, have baffled the pedantry and stupidity of such commentators as Waite. English has hardly a word to express the mental condition of such unfortunates. *Dummheit*, in its strongest German sense, is about the nearest thing to it. It is as if a geographer should criticize *Gulliver's Travels* from his own particular standpoint.

When Lévi says that all that he asserts as an initiate is subordinate to his humble submissiveness as a Christian, and then not only remarks that the Bible and the Qur'án are different translations of the same book, but treats the Incarnation as an allegory, it is evident that a good deal of submission will be required. When he agrees with St. Augustine that a thing is not just because God wills it, but God wills it because it is just, he sees perfectly well that he is reducing God to a poetic image reflected from his own moral ideal of justice, and no amount of alleged orthodoxy can weigh against that statement. His very defence of the Catholic Hierarchy is a masterpiece of that peculiar form of conscious sophistry which justifies itself by reducing its conclusion to zero. One must begin with *one*, and that *one* has no particular qualities. Therefore, so long as you have an authority properly centralized it does not really matter what that authority is. In the Pope we have such an authority ready made, and it is the gravest tactical blunder to endeavour to set up an authority opposed to him. Success in doing so means war, and failure anarchy. This, however, did not prevent Lévi from ceremonially casting a papal crown to the ground and crying, 'Death to tyranny and superstition!' in the bosom of a certain secret Areopagus of which he was the most famous member.

When a man becomes a magician he looks about him for a magical weapon; and, being probably endowed with that human frailty called laziness, he hopes to find a weapon ready made. Thus we find the Christian Magus who imposed his power upon the world taking the existing worships and making a single system combining all their merits. There is no single feature in Christianity which has not been taken bodily from the worship of Isis, or of Mithras, or of Bacchus, or of Adonis, or of Osiris. In modern times again we find Allan Bennett (Bhikkhu Ananda Metteya) trying to handle Buddhism. Others again have attempted to use Freemasonry. There have been even exceptionally foolish magicians who have tried to use a sword long since rusted.

Wagner illustrates this point very clearly in *Siegfried*. The Great Sword Nothung has been broken, and it is the only weapon that can destroy the gods. The dwarf Mime tries uselessly to mend it. When Siegfried comes he makes no such error. He melts its fragments and forges a new sword. In spite of the intense labour which this costs, it is the best plan to adopt.

Lévi completely failed to capture Catholicism; and his hope of using Imperialism, his endeavour to persuade the Emperor that he was the chosen instrument of the Almighty, a belief which would have enabled him to play Maximus to little Napoleon's Julian, was shattered once for all at Sedan.

It is necessary for the reader to gain this clear conception of Lévi's inmost mind if he is to reconcile the 'contradictions' which leave Waite petulant and bewildered. It is the sad privilege of the higher order of mind to be able to see both sides of every question, and to appreciate the fact that both are equally tenable. Such contradictions can, of course, only be reconciled on a higher plane, and this method of harmonizing contradictions is, therefore, the best key to the higher planes.

It seems unnecessary to add anything to these few remarks. This is the only difficulty in the whole book, though in one or two passages Lévi's extraordinarily keen sense of humour leads him to indulge in a little harmless bombast. We may instance his remarks on the *Grimoire* of Honorius.

We have said that this is the masterpiece of Lévi. He reaches an exaltation of both thought and language which is equal to that of any other writer known to us.

A.C.

Preface

On the brink of mystery, the spirit of man is seized with giddiness. Mystery is the abyss which ceaselessly attracts our unquiet curiosity by the terror of its depth.

The greatest mystery of the infinite is the existence of Him for whom alone all is without mystery.

Comprehending the infinite which is essentially incomprehensible, He is Himself that infinite and eternally unfathomable mystery; that is to say, that He is, in all seeming, that supreme absurdity in which Tertullian believed.

Necessarily absurd, since reason must renounce for ever the project of attaining to Him; necessarily credible, since science and reason, far from demonstrating that He does not exist, are dragged by the chariot of fatality to believe that He does exist, and to adore Him themselves with closed eyes.

Why? Because this absurd is the infinite source of reason. The light springs eternally from the eternal shadows. Science, that Babel Tower of the spirit, may twist and coil its spirals ever-ascending as it will; it may make the earth tremble, it will never touch the sky.

God is He whom we shall eternally learn to know better, and, consequently, He whom we shall never know entirely.

The realm of mystery is, then, a field open to the conquests of the intelligence. March there as boldly as you will, never will you diminish its extent; you will only alter its horizons. To know all is an impossible dream; but woe unto him who dares not to learn all, and who does not know that, in order to know anything, one must learn eternally!

They say that in order to learn anything well, one must forget it several times. The world has followed this method. Everything which is today debatable had been solved by the ancients. Before our annals began, their solutions, written in hieroglyphs, had already no longer any meaning for us. A man has rediscovered their key; he has opened the cemeteries of ancient science, and he gives to his century a whole world of forgotten theorems, of syntheses as simple and sublime as Nature, radiating always from unity, and multiplying themselves like numbers with proportions so exact, that the known demonstrates and reveals the unknown. To understand this science, is to see God. The author of this book, as he finishes his work, will think that he has demonstrated it.

Then, when you have seen God, the hierophant will say to you: 'Turn round!' and, in the shadow which you throw in the presence of this sun of intelligences, there will appear to you the devil, that black phantom which you see when your gaze is not fixed upon God, and when you think that your shadow fills the sky—for the vapours of the earth, the higher they go, seem to magnify it more and more.

To harmonize in the category of religion science with revelation and reason with faith, to demonstrate in philosophy the absolute principles which reconcile all the antinomies, and finally to reveal the universal equilibrium of natural forces, is the triple object of this work, which will consequently be divided into three parts.

We shall exhibit true religion with such characters, that no one, believer or unbeliever, can fail to recognize it; that will be the absolute in religion. We shall establish in philosophy the immutable characters of that Truth, which is in science, *reality*; in judgment, *reason*; and in ethics, *justice*. Finally, we shall acquaint you with the laws of Nature, whose equilibrium is stability, and we shall show how vain are the phantasies of our imagination before the fertile realities of movement and of life. We shall also invite the great poets of the future to create once more the divine comedy, no longer according to the dreams of man, but according to the mathematics of God.

Mysteries of other worlds, hidden forces, strange revelations, mysterious illnesses, exceptional faculties, spirits, apparitions, magical paradoxes, hermetic arcana, we shall say all, and we shall explain all. Who has given us this power? We do not fear to reveal it to our readers.

There exists an occult and sacred alphabet which the Hebrews attribute to Enoch, the Egyptians to Thoth or to Hermes Trismegistus, the Greeks to Cadmus and to Palamedes. This alphabet was known to the followers of Pythagoras, and is composed of absolute ideas attached to signs and numbers; by its combinations, it realizes the mathematics of thought. Solomon represented this alphabet by seventy-two names, written upon thirty-six talismans. Eastern initiates still call these the 'little keys' or clavicles of Solomon. These keys are described, and their use explained, in a book the source of whose traditional dogma is the patriarch Abraham. This book is called the

Sepher Yetzirah; with the aid of the *Sepher Yetzirah* one can penetrate the hidden sense of the Zohar, the great dogmatic treatise of the Qabalah of the Hebrews. The Clavicles of Solomon, forgotten in the course of time, and supposed lost, have been rediscovered by ourselves; without trouble we have opened all the doors of those old sanctuaries where absolute truth seemed to sleep—always young, and always beautiful, like that princess of the childish legend, who, during a century of slumber, awaits the bridegroom whose mission it is to awaken her.

After our book, there will still be mysteries, but higher and farther in the infinite depths. This publication is a light or a folly, a mystification or a monument. Read, reflect, and judge.

ÈLIPHAS LÉVI

PART I

Religious Mysteries

PROBLEMS FOR SOLUTION

I   To demonstrate in a certain and absolute manner the existence of God, and to give an idea of Him which will satisfy all minds.

II   To establish the existence of a true religion in such a way as to render it incontestable.

III   To indicate the bearing and the *raison d'être* of all the mysteries of the one true and universal religion.

IV   To turn the objections of philosophy into arguments favourable to true religion.

V   To draw the boundary between religion and superstition, and to give the reason of miracles and prodigies.

PRELIMINARY CONSIDERATIONS

WHEN Count Joseph de Maistre, that grand and passionate lover of Logic, said despairingly: 'The world is without religion,' he resembled those people who say rashly: 'There is no God.'

The world, in truth, is without the religion of Count Joseph de Maistre, as it is probable that such a God as the majority of atheists conceive does not exist.

Religion is an idea based upon one constant and universal fact; man is a religious animal. The word 'religion' has then a necessary and absolute sense. Nature herself sanctifies the idea which this word represents, and exalts it to the height of a principle.

The need of believing is closely linked with the need of loving; for that reason our souls need communion in the same hopes and in the same love. Isolated beliefs are only doubts: it is the bond of mutual confidence which, by creating faith, composes religion.

Faith does not invent itself, does not impose itself, does not establish itself by any political agreement; like life, it manifests itself with a sort of fatality. The same power which directs the phenomena of Nature, extends and limits the supernatural domain of faith, despite all human foresight. One does not imagine revelations; one undergoes them, and one believes in them. In vain does the spirit protest against the obscurities of dogma; it is subjugated by the attraction of these very obscurities, and often the least docile of reasoners would blush to accept the title of 'irreligious man'.

Religion holds a greater place among the realities of life than those who do without religion—or pretend to do without it—affect to believe. All ideas that raise man above the animal—moral love, devotion, honour—are sentiments essentially religious. The cult of the fatherland and of the family, fidelity to an oath and to memory, are things which humanity will never abjure without degrading itself utterly, and which could never exist without the belief in something greater than mortal life, with all its vicissitudes, its ignorance and its misery.

If annihilation were the result of all our aspirations to those sublime things which we feel to be eternal, our only duties would be enjoyment of the present, forgetfulness of the past, and carelessness about the future, and it would be rigorously true to say, as a celebrated sophist once said, that the man who thinks is a degraded animal.

Moreover, of all human passions, religious passion is the most powerful and the most lively. It generates itself, whether by affirmation or negation, with an equal fanaticism, some obstinately affirming the god that they have made in their own image, the others denying God with rashness, as if they had been able to understand and to lay waste by a single thought all that world of infinity which pertains to His great name.

Philosophers have not sufficiently considered the physiological fact of religion in humanity, for in truth religion exists apart from all dogmatic discussion. It is a faculty of the human soul just as much as intelligence and love. While man exists, so will religion. Considered in this light, it is nothing but the need of an infinite idealism, a need which justifies every aspiration for progress, which inspires every devotion, which alone prevents virtue and honour from being mere words, serving to exploit the vanity of the weak and the foolish to the profit of the strong and the clever.

It is to this innate need of belief that one might justly give the name of natural religion; and all which tends to clip the wings of these beliefs is, on the religious plane, in opposition to nature. The essence of the object of religion is mystery, since faith begins with the un-known, abandoning the rest to the investigations of science. Doubt is, moreover, the mortal enemy of faith; faith feels that the intervention of the divine being is necessary to fill the abyss which separates the finite from the infinite, and it affirms this intervention with all the warmth of its heart, with all the docility of its intelligence. If separated from this act of faith, the need of religion finds no satisfaction, and turns to scepticism and to despair. But in order that the act of faith should not be an act of folly, reason wishes it to be directed and ruled. By what? By science? We have seen that science can do nothing here. By the civil authority? It is absurd. Are our prayers to be superintended by policemen?

There remains, then, moral authority, which alone is able to constitute dogma and establish the discipline of worship, in concert this time with the civil authority, but not in obedience to its orders. It is necessary, in a word, that faith should give to the religious need a real satisfaction—a satisfaction entire, permanent, and indubitable. To obtain that, it is necessary to have the absolute and invariable affirmation of a dogma preserved by an authorized hierarchy. It is necessary to have an efficacious cult, giving, with an absolute faith, a substantial realization of the symbols of belief.

Religion thus understood being the only one which can satisfy the natural need of religion, it must be the only really natural religion. We arrive, without help from others, at this double definition, that true natural religion is revealed religion. The true revealed religion is the hierarchical and traditional religion, which affirms itself abso-lutely, above human discussion, by communion in faith, hope, and charity.

Representing the moral authority, and realizing it by the efficacy of its ministry, the priesthood is as holy and infallible as humanity is subject to vice and to error. The priest, *qua* priest, is always the representative of God. Of little account are the faults or even the crimes of man. When Alexander VI consecrated his bishops, it was not the poisoner who laid his hands upon them, it was the Pope. Pope Alexander VI never corrupted or falsified the dogmas which con-demned him, or the sacraments which in his hands saved others, and did not justify him. At all times and in all places there have been liars and criminals, but in the hierarchical and divinely authorized Church there have never been, and there will never be, either bad popes or bad priests. 'Bad' and 'priest' form an oxymoron.

We have mentioned Alexander VI, and we think that this name will be sufficient without other memories as justly execrated as his being brought up against us. Great criminals have been able to dishonour themselves doubly because of the sacred character with which they were invested, but they had not the power to dishonour that character, which remains always radiant and splendid above fallen humanity.[1]

We have said that there is no religion without mysteries; let us add that there are no mysteries without symbols. The symbol, being the formula or the expression of the mystery, only expresses its unknown depth by paradoxical images borrowed from the known. The symbolic form, having for its object to characterize what is above scientific reason, should necessarily find itself without that reason: hence the celebrated and perfectly just remark of a Father of the Church: 'I believe because it is absurd. *Credo quia absurdum.*'

If science were to affirm what it did not know, it would destroy itself. Science will then never be able to perform the work of faith, any more than faith can decide in a matter of science. An affirmation of faith with which science is rash enough to meddle can then be nothing but an absurdity for it, just as a scientific statement, if given us as an article of faith, would be an absurdity on the religious plane. To know and to believe are two terms which can never be confounded.

It would be equally impossible to oppose the one to the other. It is impossible, in fact, to believe the contrary of what one knows without ceasing, for that very reason, to know it; and it is equally impossible to achieve a knowledge contrary to what one believes without ceasing immediately to believe.

To deny or even to contest the decisions of faith in the name of science is to prove that one understands neither science nor faith: in fine, the mystery of a God of three persons is not a problem of mathematics; the incarnation of the Word is not a phenomenon in obstetrics; the scheme of redemption stands apart from the criticism of the historian. Science is absolutely powerless to decide whether we are right or wrong in believing or disbelieving dogma; it can only observe the results of belief, and if faith evidently improves men, if, moreover, faith is in itself considered as a physiological fact, evidently a necessity and a force, science will certainly be obliged to admit it, and take the wise part of always reckoning with it.

[1] A dog has six legs. Definition. It is no answer to this to show that all dogs have four.—A.C.

Let us now dare to affirm that there exists an immense fact equally appreciable both by faith and science; a fact which makes God visible (in a sense) upon earth; a fact incontestable and of universal bearing; this fact is the manifestation in the world, beginning from the epoch when the Christian revelation was made, of a spirit unknown to the ancients, of a spirit evidently divine, more positive than science in its works, in its aspirations, more magnificently ideal than the highest poetry, a spirit for which it was necessary to create a new name, a name altogether unheard[1] in the sanctuaries of antiquity. This name was created, and we shall demonstrate that this name, this word, is, in religion, as much for science as for faith, the expression of the absolute. The word is CHARITY, and the spirit of which we speak is *the Spirit of Charity*.

Before Charity, faith prostrates itself, and conquered science bows. There is here evidently something greater than humanity; Charity proves by its works that it is not a dream. It is stronger than all the passions; it triumphs over suffering and over death; it makes God understood by every heart, and seems already to fill eternity by the begun realization of its legitimate hopes.

Before Charity alive and in action who is the Proudhon who dares blaspheme? Who is the Voltaire who dares laugh?

Pile one upon the other the sophisms of Diderot, the critical arguments of Strauss, the 'Ruins' of Volney, so well named, for this man could make nothing but 'ruins', the blasphemies of the revolution whose voice was extinguished once in blood, and once again in the silence of contempt; join to it all that the future may hold for us of monstrosities and of vain dreams; then will there come the humblest and the simplest of all Sisters of Charity—the world will leave there all its follies, and all its crimes and all its dreams, to bow before this sublime reality.

Charity! word divine, sole word which makes God understood, word which contains a universal revelation! *Spirit* of *Charity*, alliance of two words which are a complete solution and a complete promise! To what question, in fine, do these two words not find an answer?

What is God for us, if not the Spirit of Charity? What is orthodoxy? Is it not the Spirit of Charity which refuses to discuss faith lest it should trouble the confidence of simple souls, and disturb the peace

---

[1] Who, however, had the word laid aside against the time when Paul should give it a meaning.—A.C.

of universal communion?[1] And the universal Church, is it any other thing than a communion in the Spirit of Charity? It is by the Spirit of Charity that the Church is infallible. It is the Spirit of Charity which is the divine virtue of the priesthood.

Duty of man, guarantee of his rights, proof of his immortality, eternity of happiness commencing for him upon the earth, glorious aim given to his existence, goal and path of all his struggles, perfection of his individual, civil and religious morality, the Spirit of Charity understands all, and is able to hope all, undertake all, and accomplish all.

It is by the Spirit of Charity that Jesus expiring on the Cross gave a son to His mother in the person of St. John, and, triumphing over the anguish of the most frightful torture, gave a cry of deliverance and of salvation, saying: 'Father, into Thy hands I commend my spirit!'

It is by charity that twelve Galilean artisans conquered the world; they loved truth more than life, and they went without followers to speak it to peoples and to kings; tested by torture, they were found faithful. They showed to the multitude a living immortality in their death, and they watered the earth with a blood whose heat could not be extinguished, because they were burning with the ardours of Charity.

It is by Charity that the Apostles built up their Creed. They said that to believe together was worth more than to doubt separately; they constituted the hierarchy on the basis of obedience—rendered so noble and so great by the Spirit of Charity, that to serve in this manner is to reign; they formulated the faith of all and the hope of all, and they put this Creed in the keeping of the Charity of all. Woe to the egoist who appropriates to himself a single word of this inheritance of the Word; he is a deicide, who wishes to dismember the body of the Lord.

This Creed is the holy ark of Charity; whoso touches it is stricken by eternal death, for Charity withdraws itself from him. It is the sacred inheritance of our children, it is the price of the blood of our fathers!

It is by Charity that the martyrs took consolation in the prisons of the Caesars, and won over to their belief even their warders and their executioners.

[1] Sublime humour of sophistry! Lévi asserts: 'Any lie will serve, provided everyone acquiesces in it,' and reprehends Christianity for disturbing the peace of Paganism. *Or*, indicates that Christianity is but syncretic-eclectic Paganism, and defends it on this ground.—A.C.

It is in the name of Charity that St. Martin of Tours protested against the torture of the Priscillians,[1] and separated himself from the communion of the tyrant who wished to impose faith by the sword.

It is by Charity that so great a crowd of saints have forced the world to accept them as expiation for the crimes committed in the name of religion itself, and the scandals of the profaned sanctuary.

It is by Charity that St. Vincent de Paul and Fénelon compelled the admiration of even the most impious centuries, and quelled in advance the laughter of the children of Voltaire before the imposing dignity of their virtues.

It is by Charity, finally, that the folly of the Cross has become the wisdom of the nations, because every noble heart has understood that it is greater to believe with those who love, and who devote themselves, than to doubt with the egotists and with the slaves of pleasure.

---

[1] The Priscillianist heresy was disturbing the Church, especially in Spain. The Emperor Maximus, a Spaniard, was inclined to put it down with a strong hand and confiscate the heretics' property. The Gallic clergy hounded him on, and the Councils of Bordeaux and Saragossa encouraged him. Two Spanish priests, *Ithacus* and *Idacus*, clamoured for the heretics' punishment by the secular arm. But St. Martin of Tours, stalwart champion of orthodoxy as he was, resisted, and in 385 he went to Trèves to plead for the persecuted Priscillianists. He prevailed. So long as Martin stayed at court the Ithacan party was foiled. When he left they had the upper hand again, and Maximus gave the suppression of the heretics into the hands of the unrelenting Evodius. Priscillian was killed. Exile and death were the fate of his followers. Heresy blazed the stronger, and a worse persecution was threatened. Then St. Martin left his cell at Marmontier, and set out a second time to Trèves. News of the old man coming along the road on his ass reached his enemies. They met him at the gate and refused him entrance. 'But,' said Martin, 'I come with the peace of Jesus Christ.' And such was the power of his presence that they could not close the city gates against him. But the palace doors were closed. Martin refused to see the Ithacans or to receive the Communion with them, and their fury at this is eloquent testimony of their sense of his power. They appealed to Maximus, who delivered over Martin bound to them. But in the night Maximus sent for Martin, argued, coaxed, persuaded him to compromise. The schism would be great, he persisted, if Martin continued to exasperate the Ithacans. Martin said he had nothing to do with persecutors. In wrath the Emperor let him go, and gave orders to the Tribunes to depart to Spain and carry out a rigorous Inquisition. Then Martin returned to Maximus and bargained. Let this order be revoked, and he would receive Communion with the Ithacans next day at the election of the new Archbishop. The order was revoked, and Martin kept his word. But when he knew the cause of Humanity safe, he departed, and on his way back to Tours experienced a great agony. Why had he had dealings with the Ithacans? In a lonely place he pondered sadly. An angel spoke to him. 'Martin, you do right to be sad, but it was the only way.' Never again did he go to any council. He was wont to say with tears that if he had saved the heretics he himself had lost power over men and over demons.

They have outraged the meaning of the episode who explain Martin's protest as merely against the surrender of the Church to Secular Power. It was *lèse-humanité* of which he held the Ithacans guilty.

St. Martin of Tours was often called Martin the Thaumaturgist. He was noted for his power over animals.

ARTICLE I

THE TRUE GOD

G OD can only be defined by faith; science can neither deny nor affirm that He exists.

God is the absolute object of human faith. In the infinite, He is the supreme and creative intelligence of order. In the world, He is the Spirit of Charity.

Is the Universal Being a fatal machine which eternally grinds down intelligences by chance, or a providential intelligence which directs forces in order to ameliorate minds?

The first hypothesis is repugnant to reason; it is pessimistic and immoral.

Science and reason ought then to accept the second.

Yes, Proudhon, God is an hypothesis, but an hypothesis so necessary, that without it, all theorems become absurd or doubtful.

For initiates of the Qabalah, God is the absolute unity which creates and animates numbers.

The unity of the human intelligence demonstrates the unity of God.

The key of numbers is that of creeds, because signs are analogical figures of the harmony which proceeds from numbers.

Mathematics could never demonstrate blind fatality, because they are the expression of the exactitude which is the character of the highest reason.

Unity demonstrates the analogy of contraries; it is the foundation, the equilibrium, and the end of numbers. The act of faith starts from unity, and returns to unity.

We shall now sketch out an explanation of the Bible by the aid of numbers, for the Bible is the book of the images of God.

We shall ask numbers to give us the reason of the dogmas of eternal religion; numbers will always reply by reuniting themselves in the synthesis of unity.

The following pages are simply outlines of qabalistic hypotheses; they stand apart from faith, and we indicate them only as curiosities of research. It is no part of our task to make innovations in dogma,

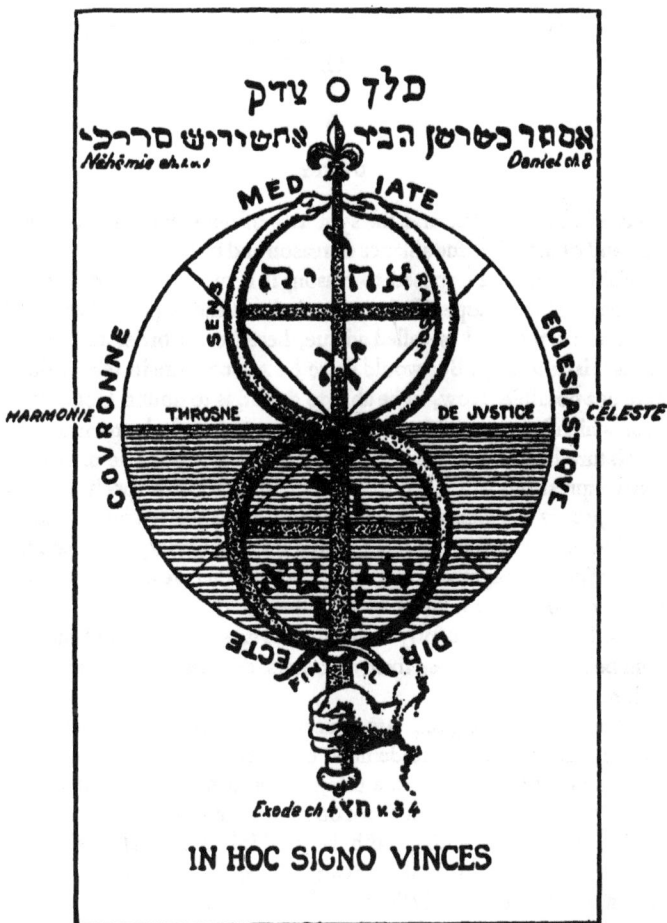

THE SIGN OF THE GRAND ARCANUM G∴ A∴

and what we assert in our character as an initiate is entirely subordinate to our submission in our character as a Christian.[1]

SKETCH OF THE PROPHETIC THEOLOGY OF NUMBERS

I

UNITY

UNITY is the principle and the synthesis of numbers; it is the idea of God and of man; it is the alliance of reason and of faith.

Faith cannot be opposed to reason; it is made necessary by love, it is identical with hope. To love is to believe and hope; and this triple outburst of the soul is called virtue, because, in order to make it, courage is necessary. But would there be any courage in that, if doubt were not possible? Now, to be able to doubt, is to doubt. Doubt is the force which balances faith, and it constitutes the whole merit of faith.

Nature herself induces us to believe; but the formulae of faith are social expressions of the tendencies of faith at a given epoch. It is that which proves the Church to be infallible, evidentially and in fact.

God is necessarily the most unknown of all beings because He is only defined by negative experience; He is all that we are not, He is the infinite opposed to the finite by hypothesis.

Faith, and consequently hope and love, are so free that man, far from being able to impose them on others, does not even impose them on himself.

'These,' says religion, 'are graces.' Now, is it conceivable that grace should be subject to demand or exaction; that is to say, could anyone wish to force men to a thing which comes freely and without price from Heaven? One must not do more than desire it for them.

To reason concerning faith is to think irrationally, since the object of faith is outside the universe of reason. If one asks me: 'Is there a God?' I reply: 'I believe it.' 'But are you sure of it?' 'If I were sure of it, I should not believe it, I should know it.'

The formulation of faith is to agree upon the terms of the common hypothesis.

Faith begins where science ends. To enlarge the scope of science

[1] This passage is typical of the sublime irony of Lévi, and the key to the whole of his paradoxes.—A.C.

is apparently to diminish that of faith; but in reality, it is to enlarge it in equal proportion, for it is to amplify its base.

One can only define the unknown by its supposed and supposable relations with the known.

Analogy was the sole dogma of the ancient magi. This dogma may indeed be called 'mediator', for it is half scientific, half hypothetical; half reason, and half poetry. This dogma has been, and will always be, the father of all others.

What is the Man-God? He who realizes, in the most human life, the most divine ideal.

Faith is a divination of intelligence and of love, when these are directed by the pointings of nature and of reason.

It is then of the essence of the things of faith to be inaccessible to science, doubtful for philosophy, and undefined for certainty.

Faith is an hypothetical realization and a conventional determination of the last aims of hope. It is the attachment to the visible sign of the things which one does not see.

'Faith is the substance of things hoped for, the evidence of things not seen.'

To affirm without folly that God is or that He is not, one must begin with a reasonable or unreasonable definition of God. Now, this definition, in order to be reasonable, must be hypothetical, analogical, and the negation of the known finite. It is possible to deny a particular god, but the absolute God can no more be denied than He can be proved; He is a reasonable supposition in whom one believes.

'Blessed are the pure in heart, for they shall see God,' said the Master; to see with the heart is to believe; and if this faith is attached to the true good, it can never be deceived, provided that it does not seek to define too much in accordance with the dangerous inductions which spring from personal ignorance. Our judgments in questions of faith apply to ourselves; it will be done to us as we have believed; that is to say, we create ourselves in the image of our ideal.

'Those who make their gods become like unto them,' says the psalmist, 'and all they that put their trust in them.'

The divine ideal of the ancient world made the civilization which came to an end, and one must not despair of seeing the god of our barbarous fathers become the devil of our more enlightened children. One makes devils with cast-off gods, and Satan is only so incoherent and so formless because he is made up of all the rags of ancient theogonies. He is the sphinx without a secret, the riddle without an answer, the mystery without truth, the absolute without reality and without light.

Man is the son of God because God, manifested, realized, and incarnated upon earth, called Himself the Son of man.

It is after having made God in the image of His intelligence and of His love, that humanity has understood the sublime Word who said 'Let there be light!'

Man is the form of the divine thought, and God is the idealized synthesis of human thought.

Thus the Word of God reveals man, and the Word of man reveals God.

Man is the God of the world, and God is the man of Heaven.

Before saying 'God wills', man has willed.

In order to understand and honour Almighty God, man must first be free.

Had he obeyed and abstained from the fruit of the tree of knowledge through fear, man would have been innocent and stupid as the lamb, sceptical and rebellious as the angel of light. He himself cut the umbilical cord of his simplicity, and, falling free upon the earth, dragged God with him in his fall.

And therefore, from this sublime fall, he rises again glorious, with the great convict of Calvary, and enters with Him into the Kingdom of Heaven.

For the Kingdom of Heaven belongs to intelligence and love, both children of liberty.

God has shown liberty to man in the image of a lovely woman, and in order to test his courage, He made the phantom of death pass between her and him.

Man loved, and felt himself to be God; he gave for her what God had just bestowed upon him—eternal hope.

He leapt towards his bride across the shadow of death.

Man possessed liberty; he had embraced life.

Expiate now thy glory, O Prometheus!

Thy heart, ceaselessly devoured, cannot die; it is thy vulture, it is Jupiter, who will die!

One day we shall awake at last from the painful dreams of a tormented life; our ordeal will be finished, and we shall be sufficiently strong against sorrow to be immortal.

Then we shall live in God with a more abundant life, and we

shall descend into His works with the light of His thought, we shall be borne away into the infinite by the whisper of His love.

We shall be without doubt the elder brethren of a new race, the angels of posterity.

Celestial messengers, we shall wander in immensity, and the stars will be our gleaming ships.

We shall transform ourselves into sweet visions to calm weeping eyes; we shall gather radiant lilies in unknown meadows, and we shall scatter their dew upon the earth.

We shall touch the eyelid of the sleeping child, and rejoice the heart of its mother with the spectacle of the beauty of her well-beloved son!

II

THE BINARY

The binary is more particularly the number of woman, mate of man and mother of society.

Man is love in intelligence; woman is intelligence in love.

Woman is the smile of the Creator content with Himself, and it is after making her that He rested, says the divine parable.

Woman stands before man because she is mother, and all is forgiven her in advance, because she brings forth in sorrow.

Woman initiated herself first into immortality through death; then man saw her to be so beautiful, and understood her to be so generous, that he refused to survive her, and loved her more than his life, more than his eternal happiness.

Happy outlaw, since she has been given to him as companion in his exile!

But the children of Cain have revolted against the mother of Abel; they have enslaved their mother.

The beauty of woman has become a prey for the brutality of such men as cannot love.

Thus woman closed her heart as if it were a secret sanctuary, and said to men unworthy of her: 'I am virgin, but I will to become mother, and my son will teach you to love me.'

O Eve! Salutation and adoration in thy fall!

O Mary! Blessings and adoration in thy sufferings and in thy glory!

Crucified and holy one who didst survive thy God that thou mightst bury thy son, be thou for us the final word of the divine revelation!

Moses called God 'Lord'; Jesus called Him 'My Father', and we, thinking of thee, may say to Providence, 'You are our mother.'

Children of woman, let us forgive fallen woman !

Children of woman, let us adore regenerate woman!

Children of woman, who have slept upon her breast, been cradled in her arms, and consoled by her caresses, let us love her, and let us love each other!

III

THE TERNARY

The Ternary is the number of creation.

God creates Himself eternally, and the infinite which He fills with His works is an incessant and infinite creation.

Supreme love contemplates itself in beauty as in a mirror, and It essays all forms as adornments, for It is the lover of life.

Man also affirms himself and creates himself; he adorns himself with his trophies of victory, he enlightens himself with his own conceptions, he clothes himself with his works as with a wedding garment.

The great week of creation has been imitated by human genius, divining the forms of nature.

Every day has furnished a new revelation, every new king of the world has been for a day the image and the incarnation of God! Sublime dream which explains the mysteries of India, and justifies all symbolisms!

The lofty conception of the man-God corresponds to the creation of Adam, and Christianity, like the first days of man in the earthly paradise, has been only an aspiration and a widowhood.

We wait for the worship of the bride and of the mother; we shall aspire to the wedding of the New Covenant.

Then the poor, the blind, the outlaws of the old world will be invited to the feast, and will receive a wedding garment. They will gaze the one upon the other with inexpressible tenderness and a smile that is ineffable because they have wept so long.

THE QUATERNARY

The Quaternary is the number of force. It is the ternary completed by its product, the rebellious unity reconciled to the sovereign trinity.

In the first fury of life, man, having forgotten his mother, no longer understood God but as an inflexible and jealous father.

The sombre Saturn, armed with his parricidal scythe, set himself to devour his children.

Jupiter had eyebrows which shook Olympus; Jehovah wielded thunders which deafened the solitudes of Sinai.

Nevertheless, the father of men, being on occasion drunken like Noah, let the world perceive the mysteries of life.

Psyche, made divine by her torments, became the bride of Eros; Adonis, raised from death, found again his Venus in Olympus; Job, victorious over evil, recovered more than he had lost.

The law is a test of courage.

To love life more than one fears the menaces of death is to merit life.

The elect are those who dare; woe to the timid!

Thus the slaves of law, who make themselves the tyrants of conscience and the servants of fear, and those who begrudge that man should hope, and the Pharisees of all the synagogues and of all the churches, are those who receive the reproofs and the curses of the Father.

Was not the Christ excommunicated and crucified by the synagogue?

Was not Savonarola burned by the order of the sovereign pontiff of the Christian religion?

Are not the Pharisees today just what they were in the time of Caiaphas?

If anyone speaks to them in the name of intelligence and love, will they listen?

In rescuing the children of liberty from the tyranny of the Pharaohs, Moses inaugurated the reign of the Father.

In breaking the insupportable yoke of Mosaic Pharisaism, Jesus welcomed all men to the brotherhood of the only Son of God.

When the last ideals fall, when the last material chains of

conscience break, when the last of them that killed the prophets and the last of them that stifled the Word are confounded, then will be the reign of the Holy Ghost.

Then, Glory to the Father who drowned the host of Pharaoh in the Red Sea!

Glory to the Son, who tore the veil of the temple, and whose Cross, overweighing the crown of the Caesars, broke the forehead of the Caesars against the earth!

Glory to the Holy Ghost, who shall sweep from the earth by His terrible breath all the thieves and all the executioners, to make room for the banquet of the children of God!

Glory to the Holy Ghost, who has promised victory over earth and over Heaven to the angel of liberty!

The angel of liberty was born before the dawn of the first day, before even the awakening of intelligence, and God called him the morning star.

O Lucifer! Voluntarily and disdainfully thou didst detach thyself from the heaven where the sun drowned thee in his splendour, to plow with thine own rays the unworked fields of night!

Thou shinest when the sun sets, and thy sparkling gaze precedes the daybreak!

Thou fallest to rise again; thou tastest of death to understand life better!

For the ancient glories of the world, thou art the evening star; for truth renascent, the lovely star of dawn.

Liberty is not licence, for licence is tyranny.

Liberty is the guardian of duty, because it reclaims right.[1]

Lucifer, of whom the dark ages have made the genius of evil, will be truly the angel of light when, having conquered liberty at the price of infamy, he will make use of it to submit himself to eternal order, inaugurating thus the glories of voluntary obedience.

Right is only the root of duty; one must possess in order to give.

This is how a lofty and profound poetry explains the fall of the angels.

God hath given to His spirits light and life; then He said to them: 'Love!'

'What is—to love?' replied the spirits.

---

[1] Right—*droit*—a word much in evidence at the time, with no true English equivalent, save in such phrases as 'the right to work'. By itself it is only used in the plural, which will not do here, and throughout this treatise.—A.C.

'To love is to give oneself to others,' replied God. 'Those who love will suffer, but they will be loved.'

'We have the right to give nothing, and we wish to suffer nothing,' said the spirits, hating love.

'Remain in your right,' answered God, 'and let us separate! I and Mine wish to suffer and even to die, to love. It is our duty!'

The fallen angel is then he who, from the beginning, refused to love; he does not love, and that is his whole torture; he does not give, and that is his poverty; he does not suffer, and that is his nothingness; he does not die, and that is his exile.

The fallen angel is not Lucifer the light-bearer; it is Satan, who calumniated love.

To be rich is to give; to give nothing is to be poor; to live is to love; to love nothing is to be dead; to be happy is to devote oneself; to exist only for oneself is to cast away oneself, and to exile oneself in hell.

Heaven is the harmony of generous thoughts; hell is the conflict of cowardly instincts.

The man of right is Cain who kills Abel from envy; the man of duty is Abel who dies for Cain for love.

And such has been the mission of Christ, the great Abel of humanity.

It is not for right that we should dare all, it is for duty.

Duty is the expansion and the enjoyment of liberty; isolated right is the father of slavery.

Duty is devotion; right is selfishness.

Duty is sacrifice; right is theft and rapine.

Duty is love, and right is hate.

Duty is infinite life; right is eternal death.

If one must fight to conquer right, it is only to acquire the power of duty; what use have we for freedom, unless to love and to devote ourselves to God?

If one must break the law, it is when law imprisons love in fear.

'He that saveth his life shall lose it,' says the Holy Book; 'and he who consents to lose it will save it.'

Duty is love; perish every obstacle to love! Silence, ye oracles of hate! Destruction to the false gods of selfishness and fear! Shame to the slaves, the misers of love!

God loves prodigal children!

V

THE QUINARY

The Quinary is the number of religion, for it is the number of God united to that of woman.

Faith is not the stupid credulity of an awestruck ignorance.

Faith is the consciousness and the confidence of love.

Faith is the cry of reason, which persists in denying the absurd, even in the presence of the unknown.

Faith is a sentiment necessary to the soul, just as breathing is to life; it is the dignity of courage, and the reality of enthusiasm.

Faith does not consist of the affirmation of this symbol or that, but of a genuine and constant aspiration towards the truths which are veiled by all symbolisms.

If a man rejects an unworthy idea of divinity, breaks its false images, revolts against hateful idolaters, you will call him an atheist!

The authors of the persecutions in fallen Rome called the first Christians atheists, because they did not adore the idols of Caligula or of Nero.

To deny a religion, even to deny all religions rather than adhere to formulae which conscience rejects, is a courageous and sublime act of faith. Every man who suffers for his convictions is a martyr of faith.

He explains himself badly, it may be, but he prefers justice and truth to everything; do not condemn him without understanding him.

To believe in the supreme truth is not to define it, and to declare that one believes in it is to recognize that one does not know it.

The Apostle St. Paul declares all faith contained in these two things: To believe that God is, and that He rewards them who seek him.

Faith is a greater thing than all religions, because it states the articles of belief with less precision.

Any dogma constitutes but a belief, and belongs to our particular communion; faith is a sentiment which is common to the whole of humanity.

The more one discusses with the object of obtaining greater accuracy, the less one believes; every new dogma is a belief which a sect appropriates to itself, and thus, in some sort, steals from universal faith.

Let us leave sectarians to make and remake their dogmas; let us leave the superstitious to detail and formulate their superstitions. As the Master said, 'Let the dead bury their dead!' Let us believe in the indicible truth; let us believe in that Absolute which reason admits without understanding it; let us believe in what we feel without knowing it!

Let us believe in the supreme reason!

Let us believe in Infinite Love, and pity the stupidities of scholasticism and the barbarities of false religion!

O man! Tell me what thou hopest, and I will tell thee what thou art worth.

Thou dost pray, thou dost fast, thou dost keep vigil; dost thou then believe that so thou wilt escape alone, or almost alone, from the enormous ruin of mankind—devoured by a jealous God? Thou art impious, and a hypocrite.

Dost thou turn life into an orgy, and hope for the slumber of nothingness? Thou art sick, and insensate.

Art thou ready to suffer as others and for others, and hope for the salvation of all? Thou art a wise and just man.

To hope is to fear not.

To be afraid of God, what blasphemy!

The act of hope is prayer.

Prayer is the flowering of the soul in eternal wisdom and in eternal love.

It is the gaze of the spirit towards truth, and the sigh of the heart towards supreme beauty.

It is the smile of the child upon its mother.

It is the murmur of the lover, who reaches out towards the kisses of his mistress.

It is the soft joy of a loving soul as it expands in an ocean of love.

It is the sadness of the bride in the absence of the bridegroom.

It is the sigh of the traveller who thinks of his fatherland.

It is the thought of the poor man who works to support his wife and children.

Let us pray in silence; let us raise towards our unknown Father a look of confidence and of love; let us accept with faith and resignation the part which He assigns to us in the toils of life, and every throb of our hearts will be a word of prayer!

Have we need to inform God of what we ask from Him? Does not He know what is necessary for us?

If we weep, let us offer Him our tears; if we rejoice, let us turn towards Him our smile; if He smite us, let us bow the head; if He caress us, let us sleep within His arms!

Our prayer will be perfect, when we pray without knowing to whom we pray.

Prayer is not a noise which strikes the ear; it is a silence which penetrates the heart.

Soft tears come to moisten the eyes, and sighs escape like incense smoke.

One feels oneself in love, ineffably in love, with all that is beauty, truth and justice; one throbs with a new life, and one fears no more to die. For prayer is the eternal life of intelligence and love; it is the life of God upon earth.

Love one another—that is the Law and the Prophets! Meditate, and understand this word.

And when you have understood, read no more, seek no more, doubt no more—love!

Be no more wise, be no more learned—love! That is the whole doctrine of true religion; religion means Charity, and God Himself is only love.

I have already said to you, to love is to give.

The impious man is he who absorbs others.

The pious man is he who loses himself in humanity.

If the heart of man concentrate in himself the fire with which God animates it, it is a hell which devours all, and fills itself only with ashes; if he radiates it without, it becomes a tender sun of love.

Man owes himself to his family; his family owes itself to the fatherland; and the fatherland to humanity.

The egoism of man merits isolation and despair; that of the family, ruin and exile; that of the fatherland, war and invasion.

The man who isolates himself from every human love, saying, 'I will serve God,' deceives himself. For, said St. John the Apostle, if he loveth not his neighbour whom he hath seen, how shall he love God whom he hath not seen?

One must render to God that which is God's, but one must not refuse even to Caesar that which is Caesar's.

God is He who gives life; Caesar can only give death.

One must love God, and not fear Caesar; as it is written in the Holy Book: 'He that taketh the sword shall perish by the sword.'

You wish to be good? Then be just. You wish to be just? Then be free.

The vices which make man like the brute are the first enemies of his liberty.

Consider the drunkard, and tell me if this unclean brute can be called free!

The miser curses the life of his father, and, like the crow, hungers for corpses.

The goal of the ambitious man is—ruins; it is the delirium of envy! The debauchee spits upon the breast of his mother, and fills with abortions the entrails of death.

All these loveless hearts are punished by the most cruel of all tortures, hate.

Because—take it to heart!—the expiation is implicit in the sin.

The man who does evil is like an earthen pot ill-made; he will break himself: fatality wills it.

With the débris of the worlds, God makes stars; with the débris of souls He makes angels.

VI

THE SENARY

The Senary is the number of initiation by ordeal; it is the number of equilibrium, it is the hieroglyph of the knowledge of Good and Evil.

He who seeks the origin of evil, seeks the source of what is not.

Evil is the disordered appetite of good, the unfruitful attempt of an unskilful will.

Everyone possesses the fruit of his work, and poverty is only the spur to toil.

For the flock of men, suffering is like the shepherd dog, who bites the wool of the sheep to put them back in the right way.

It is because of shadow that we are able to see light; because of cold that we feel heat; because of pain that we are sensible to pleasure.

Evil is then for us the occasion and the beginning of good.

But, in the dreams of our imperfect intelligence, we accuse the work of Providence, through failing to understand it.

We resemble the ignorant person who judges the picture by the beginning of the sketch, and says, when the head is done: 'What! Has this figure no body?'

Nature remains calm, and accomplishes its work.

The ploughshare is not cruel when it tears the bosom of the earth, and the great revolutions of the world are the husbandry of God.

There is a place for everything: to savage peoples, barbarous masters; to cattle, butchers; to men, judges, and fathers.

If time could change the sheep into lions, they would eat the butchers and the shepherds.

Sheep never change because they do not instruct themselves; but peoples instruct themselves.

Shepherds and butchers of the people, you are then right to regard as your enemies those who speak to your flock!

Flocks who know yet only your shepherds, and who wish to remain ignorant of their dealings with the butchers, it is excusable that you should stone them who humiliate you and disturb you, in speaking to you of your rights.

O Christ! The authorities condemn Thee, Thy disciples deny Thee, the people curses Thee, and demands Thy murder; only Thy mother weeps for Thee, even God abandons Thee!

'Eli! Eli! lama sabachthani!'

VII

THE SEPTENARY

The Septenary is the great biblical number. It is the key of the Creation in the books of Moses and the symbol of all religion. Moses left five books, and the Law is complete in two testaments.

The Bible is not a history, it is a collection of poems, a book of allegories and images.

Adam and Eve are only the primitive types of humanity; the tempter serpent is time which tests; the Tree of Knowledge is 'right'; the expiation by toil is duty.

Cain and Abel represent the flesh and the spirit, force and intelligence, violence and harmony.

The giants are those who usurped the earth in ancient times; the flood was a great revolution.

The ark is tradition preserved in a family: religion at this period becomes a mystery and the property of the race. Ham was cursed for having revealed it.

Nimrod and Babel are the two primitive allegories of the despot, and of the universal empire which has always filled the dreams of men—a dream whose fulfilment was sought successively by the Assyrians, the Medes, the Persians, Alexander, Rome, Napoleon, the successors of Peter the Great, and always unfinished because of the dispersion of interests, symbolized by the confusion of tongues.

The universal empire could not realize itself by force, but by intelligence and love. Thus, to Nimrod, the man of savage 'right', the Bible opposed Abraham, the man of duty, who goes voluntarily into exile in order to seek liberty and strife in a strange country, which he seizes by virtue of his *Idea*.

He has a sterile wife, his thought, and a fertile slave, his force; but when force has produced its fruit, thought becomes fertile; and the son of intelligence drives into exile the child of force. The man of intelligence is submitted to rude tests; he must confirm his conquests by sacrifices. God orders him to immolate his son, that is to say, doubt ought to test dogma, and the intellectual man should be ready to sacrifice everything on the altar of supreme reason. Then God intervenes: universal reason yields to the efforts of labour, and shows herself to science; the material side of dogma is alone immolated. . . . This is the meaning of the ram caught by its horns in a thicket. The history of Abraham is, then, a symbol in the ancient manner, and contains a lofty revelation of the destinies of the human soul. Taken literally, it is an absurd and revolting story. Did not St. Augustine take literally the Golden Ass of Apuleius?

Poor great men!

The history of Isaac is another legend. Rebecca is the type of the oriental woman, laborious, hospitable, partial in her affections, shrewd and wily in her manœuvres. Jacob and Esau are again the two types of Cain and Abel; but here Abel avenges himself: the emancipated intelligence triumphs by cunning. The whole of the genius of the Jews is in the character of Jacob, the patient and laborious supplanter who yields to the wrath of Esau, becomes rich, and buys his brother's forgiveness. One must never forget that, when the ancients want to philosophize, they tell a story.

The history or legend of Joseph contains, in germ, the whole genius of the Gospel; and the Christ, misunderstood by His people, must often have wept in reading over again that scene, where the Governor of Egypt throws himself on the neck of Benjamin, with the great cry of 'I am Joseph!'

Israel becomes the people of God, that is to say, the conservator of the idea, and the depository of the word. This idea is that of human independence, and of royalty, by means of work; but one hides it with care, like a precious seed. A painful and indelible sign is imprinted on the initiates; every image of the truth is forbidden, and the children of Jacob watch, sword in hand, around the unity of the tabernacle. Hamor and Shechem wish to introduce themselves forcibly into the Holy Family, and perish with their people after undergoing a feigned initiation. In order to dominate the vulgar, it is already necessary that the sanctuary should surround itself with sacrifices and with terror.

The servitude of the children of Jacob paves the way for their deliverance: for they have an idea, and one does not enchain an idea; they have a religion, and one does not violate a religion; they are, in fine, a people, and one does not enchain a real people. Persecution stirs up avengers; the idea incarnates itself in a man; Moses springs up; Pharaoh falls; and the column of smoke and flame, which goes before a freed people, advances majestically into the desert.

Christ is priest and king by intelligence and by love.

He has received the holy unction, the unction of genius, faith and virtue, which is force.

He comes when the priesthood is worn out, when the old symbols have no more virtue, when the beacon of intelligence is extinguished.

He comes to recall Israel to life, and if he cannot galvanize Israel, slain by the Pharisees, into life, he will resurrect the world given over to the dead worship of idols.

Christ is the right to do one's duty.

Man has the right to do his duty, and he has no other right.

O Man! thou hast the right to resist even unto death any who prevents thee from doing thy duty.

Mother! Thy child is drowning; a man prevents thee from helping him; thou strikest this man, thou dost run to save thy son! ... Who, then, will dare to condemn thee?

Christ came to oppose the right of duty to the duty of right.

'Right', with the Jews, was the doctrine of the Pharisees. And, indeed, they seemed to have acquired the privilege of dogmatizing; were they not the legitimate heirs of the synagogue?

They had the right to condemn the Saviour, and the Saviour knew that His duty was to resist them.

Christ is the soul of protest.

But the protest of what? Of the flesh against the intelligence? No! Of right against duty? No!

Of the physical against the moral? No! No!

Of imagination against universal reason? Of folly against wisdom? No, a thousand times No, and once more No!

Christ is the reality, duty, which protests eternally against the ideality, right.

He is the emancipation of the spirit which breaks the slavery of the flesh.

He is devotion in revolt against egoism.

He is the sublime modesty which replies to pride: 'I will not obey thee!'

Christ is unmated; Christ is solitary; Christ is sad. Why?

Because woman has prostituted herself.

Because society is guilty of theft.

Because selfish joy is impious.

Christ is judged, condemned, and executed; and men adore Him!

This happened in a world perhaps as serious as our own.

Judges of the world in which we live, pay attention, and think of Him who will judge your judgments!

But, before dying, the Saviour bequeathed to His children the immortal sign of salvation, Communion.

Communion! Common union! the final word of the Saviour of the world!

'The Bread and the Wine shared among all,' said He, 'this is my flesh and my blood.'

He gave His flesh to the executioners, His blood to the earth which drank it. Why?

In order that all may partake of the bread of intelligence, and of the wine of love.

O sign of the union of men! O Round Table of universal chivalry! O banquet of fraternity and equality! When will you be better understood?

Martyrs of humanity, all ye who have given your life in order that all should have the bread which nourishes and the wine which fortifies, do ye not also say, placing your hands on the signs of the universal communion: 'This is our flesh and our blood'?

And you, men of the whole world, you whom the Master calls His brothers; oh, do you not feel that the universal bread, the fraternal bread, the bread of the communion, is God?

Retailers of the Crucified One!

All you who are not ready to give your blood, your flesh and, your life to humanity, you are not worthy of the Communion of the Son of God! Do not let His blood flow upon you, for it would brand your forehead!

Do not approach your lips to the heart of God, He would feel your sting!

Do not drink the blood of the Christ, it will burn your entrails; it is quite sufficient that it should have flowed uselessly for you!

VIII

THE NUMBER EIGHT

The Ogdoad is the number of reaction and of equilibrating justice.

Every action produces a reaction.

This is the universal law of the world.

Christianity must needs produce anti-Christianity.

Antichrist is the shadow, the foil, the proof of Christ.

Antichrist already produced itself in the Church in the time of the Apostles: St. Paul said: 'For the mystery of iniquity doth already work; only he who now letteth will let, until he be taken out of the way. And then shall that Wicked One be revealed. . . .'[1]

The Protestants said: 'Antichrist is the Pope.'

The Pope replied: 'Every heretic is an Antichrist.'

The Antichrist is no more the Pope than Luther; the Antichrist is the spirit opposed to that of Christ.

It is the usurpation of right for the sake of right; it is the pride of domination and the despotism of thought.

It is the selfishness, self-styled religious, of Protestants, as well as the credulous and imperious ignorance of bad Catholics.

The Antichrist is what divides men instead of uniting them; it is a spirit of dispute, the obstinacy of theologians and sectarians, the impious desire of appropriating the truth to oneself, and excluding others from it, or of forcing the whole world to submit to the narrow yoke of our judgments.

The Antichrist is the priest who curses instead of blessing, who

---

[1] 2 Thess. ii. 7, 8. This passage is presumably that referred to by the author. *Cf.* 1 John iv. 3, and ii. 18.—A.C.

drives away instead of attracting, who scandalizes instead of edifying, who damns instead of saving.

It is the hateful fanaticism which discourages good-will.

It is the worship of death, sadness, and ugliness.

'What career shall we choose for our son?' have said many stupid parents; 'he is mentally and bodily weak, and he is without a spark of courage: we will make a priest of him, so that he may "live by the altar".' They have not understood that the altar is not a manger for slothful animals.

Look at the unworthy priests, contemplate these pretended servants of the altar! What do they say to your heart, these obese or cadaverous men with the lack-lustre eyes, and pinched or gaping mouths?[1]

Hear them talk: what does it teach you, their disagreeable and monotonous noise?

They pray as they sleep, and they sacrifice as they eat.

They are machines full of bread, meat and wine, and of sense-less words.

And when they plume themselves, like the oyster in the sun, on being without thought and without love, one says that they have peace of soul!

They have the peace of the brute. For man, that of the tomb is better: these are the priests of folly and ignorance, these are the ministers of Antichrist.

The true priest of Christ is a man who lives, suffers, loves and fights for justice. He does not dispute, he does not reprove; he sends out pardon, intelligence and love.

The true Christian is a stranger to the sectarian spirit; he is all things to all men, and looks on all men as the children of a common father, who means to save them all. The whole cult has for him only a sense of sweetness and of love; he leaves to God the secrets of justice, and understands only charity.

He looks on the wicked as invalids whom one must pity and cure; the world, with its errors and vices, is to him God's hospital, and he wishes to serve in it.

He does not think that he is better than anyone else; he says only: 'So long as I am in good health, let me serve others; and when I must fall and die, perhaps others will take my place and serve.'

---

[1] Actual priests. Lévi's ideal priest is in that 'Church' which is also Ark, Rose, Font, Altar, Cup, and the rest. He is that Word of Truth which is 'established' by two witnesses.—A.C.

IX

THE NUMBER NINE

This is the hermit of the Tarot; the number which refers to initiates and to prophets.

The prophets are solitaries, for it is their fate that none should ever hear them.

They see differently from others; they forefeel misfortunes. So, people imprison them and kill them, or mock then, repulse them as if they were lepers, and leave them to die of hunger.

Then, when the predictions come true, they say, 'It is these people who have brought us misfortune.'

Now, as is always the case on the eve of great disasters,[1] our streets are full of prophets.

I have met some of them in the prisons, I have seen others who were dying forgotten in garrets.

The whole great city has seen one of them whose silent prophecy was to turn ceaselessly as he walked, covered with rags, in the palace of luxury and riches.

I have seen one of them whose face shone like that of Christ: he had callosities on his hands, and wore the workman's blouse; with clay he kneaded epics. He twisted together the sword of right and the sceptre of duty; and upon this column of gold and steel he placed the creative sign of love.

One day, in a great popular assembly, he went down into the road with a piece of bread in his hand which he broke and distributed, saying: 'Bread of God, do thou make bread for all!'

I know another of them who cried: 'I will no longer adore the god of the devil! I will not have a hangman for my God!' And they thought that he blasphemed.

No; but the energy of his faith overflowed in inexact and imprudent words.

He said again in the madness of his wounded charity: 'The liabilites of all men are common, and they expiate each other's faults, as they make merit for each other by their virtues.

---

[1] This is the true clairvoyant Lévi. The Lévi who prophesied Universal Empire for Napoleon III was either the Magus trying to use him as a tool, or a Micaiah unadjured.—A.C.

'The penalty of sin is death.

'Sin itself, moreover, is a penalty, and the greatest of penalties. A great crime is nothing but a great misfortune.

'The worst of men is he who thinks himself better than his fellows.

'Passionate men are excusable, because they are passive; passion means suffering, and also redemption through sorrow.

'What we call liberty is nothing but the all-mightiness of divine compulsion. The martyrs said: "It is better to obey God than man".

'The least perfect act of love is worth more than the best act of piety.

'Judge not; speak hardly at all; love and act.'

Another prophet came and said: 'Protest against bad doctrines by good works, but do not separate yourselves.

'Rebuild all the altars, purify all the temples, and hold yourselves in readiness for the visit of the Spirit.

'Let every one pray in his own fashion, and hold communion with his own; but do not condemn others.

'A religious practice is never contemptible, for it is the sign of a great and holy thought.

'To pray together is to communicate in the same hope, the same faith, and the same charity.

'The sign by itself is nothing; it is the faith which sanctifies it.

'Religion is the most sacred and the strongest bond of human association, and to perform an act of religion is to perform an act of humanity.'

When men understand at last that one must not dispute about things about which one is ignorant,

When they feel that a little charity is worth more than much influence and domination,

When the whole world respects what even God respects in the least of His creatures, the spontaneity of obedience and the liberty of duty,

Then there will be no more than one religion in the world, the Christian and universal religion, the true Catholic religion, which will no longer deny itself by restrictions of place and of persons.

'Woman,' said the Saviour to the woman of Samaria, 'Verily I say unto thee, that the time cometh when men shall no longer worship God, either in Jerusalem, or on this mountain; for God is a spirit,[1] and they that worship Him must worship Him in spirit and in truth.'

---

[1] A mistranslation by monotheists. The Greek is, 'Spirit is God'.—A.C.

THE ABSOLUTE NUMBER OF THE QABALAH

The key of the Sephiroth. (Vide *Transcendental Magic*.)

XI

THE NUMBER ELEVEN

Eleven is the number of force; it is that of strife and martyrdom.

Every man who dies for an idea is a martyr, for in him the aspirations of the spirit have triumphed over the fears of the animal.

Every man who falls in war is a martyr, for he dies for others.

Every man who dies of starvation is a martyr, for he is like a soldier struck down in the battle of life.

Those who die in defence of right are as holy in their sacrifice as the victims of duty, and in the great struggles and revolutions against power, martyrs fell equally on both sides.

Right being the root of duty, our duty is to defend our rights.

What is a crime? The exaggeration of a right. Murder and theft are the negations of society; it is the isolated despotism of an individual who usurps royalty, and makes war at his own risk and peril.

Crime should doubtless be repressed, and society must defend itself; but who is so just, so great, so pure, as to pretend that he has the right to punish?

Peace then to all who fall in war, even in unlawful war! For they have staked their heads and they have lost them; they have paid, and what more can we ask of them?

Honour to all those who fight bravely and loyally! Shame only on the traitors and cowards!

Christ died between two thieves, and He took one of them with Him to Heaven.

The Kingdom of Heaven suffereth violence, and the violent take it by force.

God bestows His almighty power on love. He loves to triumph over hate, but the lukewarm He spueth forth from His mouth.

Duty is to live, were it but for an instant!

It is fine to have reigned for a day, even for an hour! though it were beneath the sword of Damocles, or upon the pyre of Sardanapalus!

But it is finer to have seen at one's feet all the crowns of the world, and to have said: 'I will be the king of the poor, and my throne shall be on Calvary.'

There is one man stronger than the man that slays; it is he who dies to save others.

There are no isolated crimes and no solitary expiations. There are no personal virtues, nor are there any wasted devotions.

Whoever is not without reproach is the accomplice of all evil; and whoever is not absolutely perverse, may participate in all good.

For this reason an agony is always an humanitarian expiation, and every head that falls upon the scaffold may be honoured and praised as the head of a martyr.

For this reason also the noblest and the holiest of martyrs could inquire of his own conscience, find himself deserving of the penalty that he was about to undergo, and say, saluting the sword that was ready to strike him: 'Let justice be done!'

Pure victims of the Roman Catacombs, Jews and Protestants massacred by unworthy Christians!

Priests of l'Abbaye and les Carmes,[1] victims of the Reign of Terror, butchered royalists, revolutionaries sacrificed in your turn, soldiers of our great armies who have sown the world with your bones, all you who have suffered the penalty of death, workers, strivers, darers of every kind, brave children of Prometheus, who have feared neither the lightning nor the vulture, all honour to your scattered ashes! Peace and veneration to your memories! You are the heroes of progress, martyrs of humanity!

XII

THE NUMBER TWELVE

Twelve is the cyclic number; it is that of the universal Creed.

Here is a translation in alexandrines of the unrestricted magical and Catholic creed:

[1] Monasteries in Paris which were used as prisons in the Reign of Terror.—A.C.

I do believe in God, almighty sire of man,
One God, who did create the universe, His plan.

I do believe in Him, the Son, the chief of men,
Word and magnificence of the supreme Amen.

He is the living thought of Love's eternal might,
God manifest in flesh, the Action of the Light,

Desired in every place and every period,
But not a god that one may separate from God.

Descended among men to free the earth from fate,
He in His mother did the woman consecrate.

He was the man whom heaven's sweet wisdom did adorn;
To suffer and to die as men do He was born.

Proscribed by ignorance, accused by envy and strife,
He died upon the cross that He might give us life.

All who accept His aid to guide and to sustain
By His example may to God like Him attain.

He rose from death to reign through the ages' dance;
He is the sun that melts the clouds of ignorance.

His precepts, better known and mightier soon to be,
Shall judge the quick and dead for all eternity.

I do believe in God's most Holy Spirit, whose fire,
The heart and mind of saints and prophets did inspire.

He is a Breath of life and of fecundity,
Proceeding both from God and from humanity.

I do believe in one most holy brotherhood,
Of just men that revere heaven's ordinance of good.

I do believe one place, one pontiff, and one right,
One symbol of one God, in one intent unite.

I do believe that death by changing us renews,
And that in man as God life sheds immortal dews.

THE NUMBER THIRTEEN

Thirteen is the number of death and of birth; it is that of property and of inheritance, of society and of family, of war and of treaties.

The basis of society is the exchange of right, duty and good faith.

Right is property, exchange is necessity, good faith is duty.

He who wants to receive more than he gives, or who wants to receive without giving, is a thief.

Property is the right to dispose of a portion of the common wealth; it is not the right to destroy, nor the right to sequestrate.

To destroy or sequestrate the common wealth is not to possess; it is to steal.

I say common wealth, because the true proprietor of all things is God, who wishes all things to belong to everybody. Whatever you may do, at your death you will carry away nothing of this world's goods. Now, that which must be taken away from you one day is not really yours. It has only been lent to you.

As to the usufruct, it is the result of work; but even work is not an assured guarantee of possession, and war may come with devastation and fire to displace property.

Make then good use of those things which perish, O you who will perish before they do!

Consider that egoism provokes egoism, and that the immorality of the rich man will answer for the crimes of the poor.

What does the poor man wish, if he is honest? He wishes for work.

Use your rights, but do your duty: the duty of the rich man is to spread wealth; wealth which does not circulate is dead; do not hoard death!

A sophist[1] has said, 'Property is robbery,' and he doubtless wished to speak of property absorbed in itself, withdrawn from free exchange, turned from common use.

If such was his thought, he might go further, and say that such a suppression of public life is indeed assassination.

It is the crime of monopoly, which public instinct has always looked upon as treason to the human race.

The family is a natural society which results from marriage.

[1] Proudhon.—A.C.

Marriage is the union of two beings joined by love, who promise each other mutual devotion in the interest of the children who may be born.

Married persons who have a child, and who separate, are impious. Do they then wish to execute the judgment of Solomon and hew the child asunder?

To vow eternal love is puerile; sexual love is an emotion, divine doubtless, but accidental, involuntary and transitory; but the promise of reciprocal devotion is the essence of marriage and the fundamental principle of the family.

The sanction and the guarantee of this promise must then be an absolute confidence.

Every jealousy is a suspicion, and every suspicion is an outrage.

The real adultery is the breach of this trust: the woman who complains of her husband to another man; the man who confides to another woman the disappointments or the hopes of his heart—these do, indeed, betray conjugal faith.

The surprises which one's senses spring upon one are only infidelities on account of the impulses of the heart which abandons itself more or less to the whispers of pleasure. Moreover, these are human faults for which one must blush, and which one ought to hide: they are indecencies which one must avoid in advance by removing opportunity, but which one must never seek to surprise: morality proscribes scandal.

Every scandal is a turpitude. One is not indecent because one possesses organs which modesty does not name, but one is obscene when one exhibits them.

Husbands, hide your domestic wounds; do not strip your wives naked before the laughter of the mob!

Women, do not advertise the discomforts of the conjugal bed: to do so is to write yourselves prostitutes in public opinion.

It needs a lofty degree of courage to keep conjugal faith; it is a pact of heroism of which only great souls can understand the whole extent.

Marriages which break are not marriages: they are couplings.

A woman who abandons her husband, what can she become? She is no more a wife, and she is not a widow; what is she then? She is an apostate from honour who is forced to be licentious because she is neither virgin nor free.

A husband who abandons his wife prostitutes her, and deserves the infamous name that one applies to the lovers of lost women.

Marriage is then sacred and indissoluble when it really exists.

But it cannot really exist, except for beings of a lofty intelligence and of a noble heart.

The animals do not marry, and men who live like animals undergo the fatalities of the brute nature.

They ceaselessly make unfortunate attempts to act reasonably. Their promises are attempts at and imitations of promises; their marriages, attempts at and imitations of marriage; their loves, attempts at and imitations of love. They always wish, and never will; they are always undertaking and never completing. For such people, only the repressive side of law applies.

Such beings may have a litter, but they never have a family: marriage and family are the rights of the perfect man, the emancipated man, the man who is intelligent and free.

Ask also the annals of the Courts, and read the history of parricides.

Raise the black veil from off all those chopped heads, and ask them what they thought of marriage and of the family, what milk they sucked, what caresses ennobled them. . . . Then shudder, all you who do not give to your children the bread of intelligence and of love, all you who do not sanction paternal authority by the virtue of a good example!

Those wretches were orphans in spirit and in heart, and they have avenged their birth.

We live in a century when more than ever the family is misunderstood in all that it possesses which partakes of the august and the sacred: material interest is killing intelligence and love; the lessons of experience are despised, the things of God are hawked about the street. The flesh insults the spirit, fraud laughs in the face of loyalty. No more idealism, no more justice: human life has murdered both its father and its mother.

Courage and patience! This century should go where great criminals should go. Look at it, how sad it is! Weariness is the black veil of its face . . . the tumbril rolls on, and the shuddering crowd follows it. . . .

Soon one more century will be judged by history, and one will write upon a mighty tomb of ruins:

'Here ends the parricide century! The century which murdered its God and its Christ!'

In war, one has the right to kill, in order not to die: but in the battle of life the most sublime of rights is that of dying in order not to kill.

Intelligence and love should resist oppression unto death—but never unto murder.

Brave man, the life of him who has offended you is in your hands; for he is master of the life of others who cares not for his own. . . . Crush him beneath your greatness: pardon him!

'But is it forbidden to kill the tiger which threatens us?'

'If it is a tiger with a human face, it is finer to let him devour us—yet, for all that, morality has here nothing to say.'

'But if the tiger threatens my children?'

'Let Nature herself reply to you!'

Harmodius and Aristogiton had festivals and statues in Ancient Greece. The Bible has consecrated the names of Judith and Ehud, and one of the most sublime figures of the Holy Book is that of Samson, blind and chained, pulling down the columns of the temple, as he cried: 'Let me die with the Philistines!'

And yet, do you think that, if Jesus, before dying, had gone to Rome to plunge His dagger in the heart of Tiberius, He would have saved the world, as He did, in forgiving His executioners, and in dying for even Tiberius?

Did Brutus save Roman liberty by killing Caesar? In killing Caligula, Chaerea only made place for Claudius and Nero. To protest against violence by violence, is to justify it, and to force it to reproduce itself.

But to triumph over evil by good, over selfishness by self-abnegation, over ferocity by pardon, that is the secret of Christianity, and it is that of eternal victory.

I have seen the place where the earth still bled from the murder of Abel, and on that place there ran a brook of tears.

Under the guidance of the centuries, myriads of men moved on, letting fall their tears into the brook.

And Eternity, crouching mournful, gazed upon the tears which fell; she counted them one by one, and there were never enough of them to wash away one stain of blood.

But between two multitudes and two ages came the Christ, a pale and radiant figure.

And in the earth of blood and tears, He planted the vine of fraternity; and the tears and the blood, sucked up by the roots of the divine tree, became the delicious sap of the grape, which is destined to intoxicate with love the children of the future.

XIV

THE NUMBER FOURTEEN

Fourteen is the number of fusion, of association, and of universal unity, and it is in the name of what it represents that we shall here make an appeal to the nations, beginning with the most ancient and the most holy.

Children of Israel, why, in the midst of the movement of the nations, do you rest immobile, guardians of the tombs of your fathers?

Your fathers are not here, they are risen: for the God of Abraham, of Isaac and of Jacob, is not the God of the dead!

Why do you always impress upon your offspring the bloody sigil of the knife?

God no longer wishes to separate you from other men; be our brethren, and eat with us the consecrated Bread of peace on altars that blood stains never.

The law of Moses is accomplished: read your books and understand that you have been a blind and hard-hearted race, even as all your prophets said to you.

You have also been a courageous race, a race that persevered in strife.

Children of Israel, become the children of God: understand and love!

God has wiped from your forehead the brand of Cain, and the peoples seeing you pass will no longer say: 'There go the Jews!' They will cry: 'Room for our brethren! Room for our elders in the Faith!'

And we shall go every year to eat the Passover with you in the city of the New Jerusalem.

And we shall take our rest under your vine and under your fig-tree; for you will be once more the friend of the traveller, in memory of Abraham, of Tobias and of the angels who visited them.

And in memory of Him who said: 'He who receiveth the least of these My little ones, receiveth Me.'

For then you will no longer refuse an asylum in your house and in your heart to your brother Joseph, whom you sold to the Gentiles.

Because he has become powerful in the land of Egypt where you sought bread in the days of famine.

And he has remembered his father Jacob and Benjamin his young brother, and he pardons you your jealousy and embraces you with tears.

Children of true believers, we will sing with you: 'There is no God but God, and Mohammed is His prophet!'

Say with the children of Israel: 'There is no God but God, and Moses is His prophet!'

Say with the Christians: 'There is no God but God, and Jesus Christ is His prophet!'

Mohammed is the shadow of Moses. Moses is the forerunner of Jesus.

What is a prophet? A representative of humanity seeking God. God is God, and man is the prophet of God, when he causes us to believe in God.

The Old Testament, the Qur'án, and the Gospel are three different translations of the same book. As God is one, so also is the law.

O ideal woman! O reward of the elect! Art thou more beautiful than Mary?

O Mary, daughter of the East! chaste as pure love, great as the desire of motherhood, come and teach the children of Islam the mysteries of Paradise and the secrets of beauty.

Invite them to the festival of the new alliance! There, upon three thrones glittering with precious stones, three prophets will be seated.

The tuba-tree will make, with its back-curving branches, a dais for the celestial table.

The bride will be white as the moon, and scarlet as the smile of morning.

All nations shall press forward to see her, and they will no longer fear to pass Al Sirah; for, on that razor-edged bridge, the Saviour will stretch His Cross, and come to stretch His hand to those who stumble, and to those who have fallen the bride will stretch her perfumed veil and draw them to her.

O ye people, clap your hands, and praise the last triumph of love! Death alone will remain dead, and hell alone will be consumed!

O nations of Europe, to whom the East stretches forth its hands, unite and push back the northern bear![1] Let the last war bring the

---

[1] Written about the time of the Crimean War, this indicates Lévi's attempt to use Imperialism as his magical weapon, just as Allan Bennett tried to use Buddhism. All these second-hand swords break, as Wagner saw when he wrote *Siegfried*, and invented a new Music, a Nothung which has shorn asunder more false sceptres than Wotan's.
—A.C.

triumph of intelligence and love, let commerce interlace the arms of the world, and a new civilization, sprung from the armed Gospel, unite all the flocks of the earth under the crook of the same shepherd!

Such will be the conquests of progress, such is the end towards which the whole movement of the world is pushing us.

Progress is movement, and movement is life.

To deny progress is to affirm nothingness, and to deify death.

Progress is the only reply that reason can give to the objections which the existence of evil raises.

All is not well, but all will be well one day. God begins His work, and He will finish it.

Without progress, evil would be immutable like God.

Progress explains ruins, and consoles the weeping of Jeremiah.

Nations succeed each other like men; and nothing is stable, because everything is marching towards perfection.

The great man who dies bequeathes to his country the fruit of his works; the great nation which becomes extinguished upon earth transforms itself into a star to enlighten the obscurities of History.

What it has written by its actions remains graven in the eternal book; it has added a page to the Bible of the human race.

Do not say that civilization is bad; for it resembles the damp heat which ripens the harvest, it rapidly develops the principles of life and the principles of death, it kills and it vivifies.

It is like the angel of the judgment who separates the wicked from the good.

Civilization transforms men of good-will into angels of light, and lowers the selfish man beneath the brute; it is the corruption of bodies and the emancipation of souls.

The impious world of the giants raised to Heaven the soul of Enoch; above the Bacchanals of primitive Greece rises the harmonious spirit of Orpheus.

Socrates and Pythagoras, Plato and Aristotle, resume, in explaining them, all the aspirations and all the glories of the ancient world; the fables of Homer remain truer than history, and nothing remains to us of the grandeur of Rome but the immortal writings which the century of Augustus brought forth.

Thus, perhaps, Rome only shook the world with the convulsions of war, in order to bring forth Vergil.

Christianity is the fruit of the meditations of all the sages of the East who live again in Jesus Christ.

Thus the light of the spirits has risen where the sun of the world rises; Christ conquered the West, and the soft rays of the sun of Asia have touched the icicles of the North.

Stirred by this unknown heat, ant-heaps of new men have spread over a worn-out world; the souls of dead people have shone upon rejuvenated races, and enlarged in them the spirit of life.

There is in the world a nation which calls itself frankness and freedom, for these two words are synonymous with the name of France.

This nation has always been in some ways more Catholic than the Pope, and more Protestant than Luther.

The France of the Crusades, the France of the Troubadours, the France of songs, the France of Rabelais and of Voltaire, the France of Bossuet and of Pascal, it is she who is the synthesis of all peoples; it is she who consecrates the alliance of reason and of faith, of revolution and of power, of the most tender belief and of the proudest human dignity.

And, see how she marches, how she swings herself, how she struggles, how she grows great!

Often deceived and wounded, never cast down, enthusiastic over her triumphs, daring in her adversities, she laughs, she sings, she dies, and she teaches the world faith in immortality.

The old guard does not surrender, but neither does it die! The proof of it is the enthusiasm of our children, who mean, one day, to be also soldiers of the old guard!

Napoleon is no more a man: he is the very genius of France, he is the second saviour of the world, and he also gave for a sign the Cross to his apostles.

St. Helena and Golgotha are the beacons of the new civilization; they are the two piles of an immense bridge made by the rainbow of the final deluge, and which throws a bridge between the two worlds.

And can you believe that a past without aureole and without glory, might capture and devour so great a future?

Could you think that the spur of a Tartar might one day tear up the pact of our glories, the testament of our liberties?

Say rather that we may again become children, and enter again into our mother's womb!

'Go on! Go on!' said the voice of God to the wandering Jew. 'Advance! Advance!' the destiny of the world cries out to France. And where do we go? To the unknown, to the abyss perhaps; no matter! But to the past, to the cemeteries of oblivion, to the swaddling-clothes which our childhood itself tore in shreds, towards the imbecility and ignorance of the earliest ages ... never! never!

XV

THE NUMBER FIFTEEN

Fifteen is the number of antagonism, and of catholicity.

Christianity is at present divided into two churches: the civilizing Church, and the savage Church; the progressive Church, and the stationary Church.

One is active, the other is passive: one has mastered the nations and governs them always, since kings fear it; the other has submitted to every despotism, and can be nothing but an instrument of slavery.

The active Church realizes God for men, and alone believes in the divinity of the human Word, as an interpreter of that of God.

What after all is the infallibility of the Pope, but the autocracy of intelligence, confirmed by the universal vote of faith?

In this case, one might say, the Pope ought to be the first genius of his century. Why? It is more proper, in reality, that he should be an average man. His supremacy is only more divine for that, because it is in a way more human.

Do not events speak louder than rancours and irreligious ignorances? Do not you see Catholic France sustaining with one hand the tottering papacy, and with the other holding the sword to fight at the head of the army of progress?

Catholics, Jews, Turks, Protestants, already fight under the same banner; the crescent has rallied to the Latin cross, and altogether we struggle against the invasion of the barbarians, and their brutalizing orthodoxy.

It is for ever an accomplished fact. In admitting new dogmas, the chair of St. Peter has solemnly proclaimed itself progressive.

The fatherland of Catholic Christianity is that of the sciences and of the fine arts; and the eternal Word of the Gospel, living and incarnate in a visible authority, is still the light of the world.

Silence, then, to the Pharisees of the new synagogue! Silence to the hateful traditions of the Schools, to the arrogance of Presbyterianism, to the absurdity of Jansenism, and to all those shameful and superstitious interpretations of the eternal dogma, so justly stigmatized by the pitiless genius of Voltaire!

Voltaire and Napoleon died Catholics.[1] And do you know what the Catholicism of the future must be?

It will be the dogma of the Gospel, tried like gold by the critical acid of Voltaire, and realized, in the kingdom of the world, by the genius of the Christian Napoleon.

Those who will not march will be dragged or trampled by events.

Immense calamities may again hang over the world. The armies of the Apocalypse may, perhaps, one day, unchain the four scourges. The sanctuary will be cleansed. Rigid and holy poverty will send forth its apostles to uphold what staggers, lift up again what is broken, and anoint all wounds with sacred oils.

Those two blood-hungered monsters, despotism and anarchy, will tear themselves to pieces, and annihilate each other, after having mutually sustained each other for a little while, by the embrace of their struggle itself.

And the government of the future will be that whose model is shown to us in nature, by the family, and in the religious world by the pastoral hierarchy. The elect shall reign with Jesus Christ during a thousand years, say the apostolic traditions: that is to say, that during a series of centuries, the intelligence and love of chosen men, devoted to the burden of power, will administer the interests and the wealth of the universal family.

At that day, according to the promise of the Gospel, there will be no more than one flock and one shepherd.

XVI

THE NUMBER SIXTEEN

Sixteen is the number of the temple.

Let us say what the temple of the future will be!

When the spirit of intelligence and love shall have revealed itself, the whole trinity will manifest itself in its truth and in its glory.

Humanity, become a queen, and, as it were, risen from the dead,

---

[1] 'I do not say that Voltaire died a good Catholic, but he died a Catholic.'—E.L. Christian authors unanimously hold that, like all 'heretics', he repented on his death-bed, and died blaspheming. What on earth does it matter? Life, not death, reveals the soul. —A.C.

will have the grace of childhood in its poesy, the vigour of youth in its reason, and the wisdom of ripe age in its works.

All those forms, which the divine thought has successively clothed, will be born again, immortal and perfect.

All those features which the art of successive nations has sketched will unite themselves, and form the complete image of God.

Jerusalem will rebuild the Temple of Jehovah on the model prophesied by Ezekiel; and the Christ, new and eternal Solomon, will chant, beneath roofs of cedar and of cypress, the Epithalamium of his marriage with holy liberty, the holy bride of the Song of Songs.

But Jehovah will have laid aside his thunderbolts, to bless with both hands the bridegroom and the bride; he will appear smiling between them, and take pleasure in being called father.

However, the poetry of the East, in its magical souvenirs, will call him still Brahma, and Jupiter. India will teach our enchanted climates the marvellous fables of Vishnu, and we shall place upon the still bleeding forehead of our well-beloved Christ the triple crown of pearls of the mystical Trimurti. From that time, Venus, purified under the veil of Mary, will no more weep for her Adonis.

The bridegroom is risen to die no more, and the infernal boar has found death in its momentary victory.

Lift yourselves up again, O Temples of Delphi and of Ephesus! The God of Light and of Art is become the God of the world, and the Word of God is indeed willing to be called Apollo! Diana will no more reign widowed in the lonely fields of night; her silvern crescent is now beneath the feet of the bride.

But Diana is not conquered by Venus; her Endymion has wakened, and virginity is about to take pride in motherhood!

Quit the tomb, O Phidias, and rejoice in the destruction of thy first Jupiter: it is now that thou wilt conceive a God!

O Rome, let thy temples rise again, side by side with thy basilicas: be once more the Queen of the World, and the Pantheon of the nations; let Vergil be crowned on the Capitol by the hand of St. Peter; and let Olympus and Carmel unite their divinities beneath the brush of Raphael!

Transfigure yourselves, ancient cathedrals of our fathers; dart forth into the clouds your chiselled and living arrows, and let stone record in animated figures the dark legends of the North, brightened by the marvellous gilded apologues of the Qur'án!

Let the East adore Jesus Christ in its mosques, and on the

minarets of a new Santa Sophia let the Cross rise in the midst of the crescent![1]

Let Mohammed set woman free to give to the true believer the houris which he has so long dreamt of, and let the martyrs of the Saviour teach chaste caresses to the beautiful angels of Mohammed!

The whole earth, reclothed with the rich adornments which all the arts have embroidered for her, will no longer be anything but a magnificent temple, of which man shall be the eternal priest.

All that was true, all that was beautiful, all that was sweet in the past centuries, will live once more glorified in this transfiguration of the world.

And the beautiful form will remain inseparable from the true idea, as the body will one day be inseparable from the soul, when the soul, come to its own power, will have made itself a body in its own image.

That will be the Kingdom of Heaven upon earth, and the body will be the temple of the soul, as the regenerated universe will be the body of God.

And bodies and souls, and form and thought, and the whole universe, will be the light, the word, and the permanent and visible revelation of God. Amen. So be it.

XVII

THE NUMBER SEVENTEEN

Seventeen is the number of the star; it is that of intelligence and love.

Warrior and bold intelligence, accomplice of divine Prometheus, eldest daughter of Lucifer, hail unto thee in thine audacity! Thou didst wish to know, and in order to possess, thou didst brave all the thunders, and affronted every abyss!

Intelligence, O Thou, whom we poor sinners have loved to madness, to scandal, to reprobation! Divine right of man, essence and soul of liberty, hail unto thee! For they have pursued thee, in trampling beneath their feet for thee the dearest dreams of their imagination, the best beloved phantoms of their heart!

For thee, they have been repulsed and proscribed, for thee they have suffered prison, nakedness, hunger, thirst, the desertion of those

[1] It is amusing to remark that this very symbol is characteristic of the Greek Church which he has been attacking. Lévi should have visited Moscow.—A.C.

whom they loved, and the dark temptations of despair! Thou wast their right, and they have conquered thee! Now they can weep and believe, now they can submit themselves and pray!

Repentant Cain would have been greater than Abel: it is lawful pride satisfied which has the right to humiliate itself!

I believe because I know why and how one must believe; I believe because I love, and fear no more.

Love! Love! Sublime redeemer and sublime restorer; thou who makest so much happiness, with so many tortures, thou who didst sacrifice blood and tears, thou who art virtue itself, and the reward of virtue; force of resignation, belief of obedience, joy of sorrow, life of death, hail! Salutation and glory to thee! If intelligence is a lamp, thou art its flame; if it is right, thou art duty; if it is nobility, thou art happiness. Love, full of pride and modesty in thy mysteries, divine love, hidden love, love insensate and sublime, Titan who takest Heaven in both hands, and forcest it to earth, final and ineffable secret of Christian widowhood, love eternal, love infinite, ideal which would suffice to create worlds; love! love! blessing and glory to thee! Glory to the intelligences which veil themselves that they may not offend weak eyes! Glory to right which transforms itself wholly into duty, and which becomes devotion! To the widowed souls who love, and burn up without being loved! To those who suffer, and make none other suffer, to those who forgive the ungrateful, to those who love their enemies! Oh, happy evermore, happy beyond all, are those who embrace poverty, who have drained themselves to the dregs, to give! Happy are the souls who for ever make thy peace! Happy the pure and the simple hearts that never think themselves better than others! Humanity, my mother, humanity daughter and mother of God, humanity conceived without sin, universal Church, Mary! Happy is he who has dared all to know thee and to understand thee, and who is ready to suffer all once more, in order to serve thee and to love thee!

XVIII

THE NUMBER EIGHTEEN

This number is that of religious dogma, which is all poetry and all mystery.

The Gospel says that at the death of the Saviour the veil of the

Temple was rent, because that death manifested the triumph of devotion, the miracle of charity, the power of God in man, divine humanity, and human divinity, the highest and most sublime of Arcana, the last word of all initiations.

But the Saviour knew that at first men would not understand him, and he said: 'You will not be able to bear at present the full light of my doctrine; but, when the Spirit of Truth shall manifest himself, he will teach you all truth, and he will cause you to understand the sense of what I have said unto you.'

Now the Spirit of Truth is the spirit of science and intelligence, the spirit of force and of counsel.

It is that spirit which solemnly manifested itself in the Roman Church, when it declared in the four articles of its decree of the 12th December, 1845:

1°. That if faith is superior to reason, reason ought to endorse the inspirations of faith;

2°. That faith and science have each their separate domain, and that the one should not usurp the functions of the other;

3°. That it is proper for faith and grace, not to weaken, but on the contrary to strengthen and develop reason;

4°. That the concourse of reason, which examines, not the decisions of faith, but the natural and rational bases of the authority which decides them, far from injuring faith, can only be useful to it; in other words, that a faith, perfectly reasonable in its principles, should not fear, but should, on the contrary, desire the sincere examination of reason.

Such a decree is the accomplishment of a complete religious revolution, it is the inauguration of the reign of the Holy Ghost upon the earth.

XIX

THE NUMBER NINETEEN

It is the number of light.

It is the existence of God proved by the very idea of God.

Either one must say that Being is the universal tomb where, by an automatic movement, stirs a form for ever dead and corpse-like, or one must admit the absolute principle of intelligence and of life.

Is the universal light dead or alive? Is it vowed fatally to the work of destruction, or providentially directed to an immortal birth?

If there be no God, intelligence is only a deception, for it fails to be the absolute, and its ideal is a lie.

Without God, being is a nothingness affirming itself, life a death in disguise, and light a night for ever deceived by the mirage of dreams.

The first and most essential act of faith is then this.

Being exists; and the Being of beings, the Truth of being, is God.

Being is alive with intelligence, and the living intelligence of absolute being is God.

Light is real and life-giving; now, the reality and life of all light is God.

The word of universal reason is an affirmation and not a negation.

How blind are they who do not see that physical light is nothing but the instrument of thought!

Thought alone, then, reveals light, and creates it in using it for its own purposes.

The affirmation of atheism is the dogma of eternal night: the affirmation of God is the dogma of light!

We stop here at the number Nineteen, although the sacred alphabet has twenty-two letters; but the first nineteen are the keys of occult theology. The others are the keys of Nature; we shall return to them in the third part of this work.

.   .   .   .   .

Let us resume what we have said concerning God, by quoting a fine invocation borrowed from the Jewish liturgy. It is a page from the qabalistic poem Kether-Malkuth, by Rabbi Solomon, son of Gabirol:

Thou art one, the beginning of all numbers, and the foundation of all buildings; thou art one, and in the secret of thy unity the most wise of men are lost, because they know it not. Thou art one, and thy unity neither wanes nor waxes, neither suffers any change. Thou art one, and yet not the one of the mathematician, for thy unity admits neither multiplication, nor change, nor form. Thou art one, and not one of mine imaginations can fix a limit for thee, or give a definition of thee; therefore will I take heed to my ways, lest I offend with my tongue. Thou art one indeed, whose excellence is so lofty, that it may in no wise fall, by no means like that one which may cease to be.

Thou art the existing one; nevertheless, the understanding and the sight of mortals cannot attain thine existence, nor place in thee the where, the how, the why. Thou art the existing one, but in thyself, since no other can exist beside thee. Thou art the existing one, before time, and beyond space. Thou art indeed the existing one, and thine existence is so hidden, and so deep, that none can discover it, or penetrate its secret.

Thou art the living one, but not in fixed and known time; thou art the living one, but not by spirit or by soul; for thou art the Soul of all souls. Thou art the living one; but not living with the life of mortals, that is, like a breath, and whose end is to give food to worms. Thou art the living one, and he that can attain thy mysteries will enjoy eternal delight and live for ever.

Thou art great; before thy greatness all other greatness bows, and all that is most excellent becomes imperfect. Thou art great above all imagination, and thou art exalted above all the hierarchies of Heaven. Thou art great above all greatness, and thou art exalted above all praise. Thou art strong, and not one among thy creatures can do the works that thou dost, nor can his force be compared with thine. Thou art strong, and it is to thee that belongs that strength invincible which changes not and decays never. Thou art strong; by thy loving-kindness thou dost forgive in the moment of thy most burning wrath, and thou showest thyself long-suffering to sinners. Thou art strong, and thy mercies, existing from all time, are upon all thy creatures. Thou art the eternal light, that pure souls shall see, and that the cloud of sins will hide from the eyes of sinners. Thou art the light which is hidden in this world, and visible in the other, where the glory of the Lord is shown forth. Thou art Sovereign, and the eyes of understanding which desire to see thee are all amazed, for they can attain but part of it, never the whole. Thou art the God of gods, and all thy creatures bear witness to it; and in honour of this great name they owe thee all their worship. Thou art God, and all created beings are thy servants and thy worshippers; thy glory is not tarnished, although men worship other gods, because their intention is to address themselves to thee; they are like blind men, who wish to follow the straight road, but stray; one falls into a well, the other into a ditch; all think that they are come to their desire, yet they have wearied themselves in vain. But thy servants are like men of clear sight travelling upon the highroad; never do they stray from it, either to the right hand or the left, until they are entered into the court of the king's palace. Thou art God, who by thy godhead sustainest all beings, and by thy unity dost bring home all creatures. Thou art God, and there is no difference between thy deity, thy unity, thy eternity, and thy existence; for all is one and the same mystery; although names vary, all returns to the same truth. Thou art the knower, and that intelligence which is the source of life emanates from thyself; and beside thy knowledge all the wisest men are fools. Thou art the knower, and thou hast learned thy knowledge from none, nor hast acquired it but from thyself. Thou art the knower, and like a workman and an architect thou hast taken from thy knowledge a divine will, at an appointed time, to draw being from nothing; so that the light which falls from the eyes is drawn from its own centre without any instrument or tool. This divine will has hollowed, designed, purified and moulded; it has ordered Nothingness to open itself, Being to shut up, and the world to spread itself. It has spanned the heavens, and assembled with its power the tabernacle of the spheres, with the cords of its might it has bound the curtains of the creatures of the universe, and touching with its strength the edge of the curtain of creation, has joined that which is above to that which was below.'—*Prayers of Kippour*

We have given to these bold qabalistic speculations the only form which suits them, that is, poesy, or the inspiration of the heart.

Believing souls will have no need of the rational hypotheses contained in this new explanation of the figures of the Bible; but those sincere hearts afflicted by doubt, which are tortured by eighteenth-century criticism, will understand in reading it that even reason without faith can find in the Holy Book something besides stumbling-blocks; if the veils with which the divine text is covered throw a great shadow, this shadow is so marvellously designed by the interplay of light that it becomes the sole intelligible image of the divine ideal.

Ideal, incomprehensible as infinity, and indispensable as the very essence of mystery!

ARTICLE II

SOLUTION OF THE SECOND PROBLEM

TRUE RELIGION

RELIGION exists in humanity, like love.

Like it, it is unique.

Like it, it either exists, or does not exist, in such and such a soul; but, whether one accepts it or denies it, it is in humanity; it is, then, in life, it is in nature itself; it is an incontestable fact of science, and even of reason.

The true religion is that which has always existed, which exists today, and will exist for ever.

Someone may say that religion is this or that; religion is what it is. This is the true religion, and the false religions are superstitions imitated from her, borrowed from her, lying shadows of herself!

One may say of religion what one says of true art. Savage attempts at painting or sculpture are the attempts of ignorance to arrive at the truth. Art proves itself by itself, is radiant with its own splendour, is unique and eternal like beauty.

The true religion is beautiful, and it is by that divine character that it imposes itself on the respect of science, and obtains the assent of reason.

Science dare not affirm or deny those dogmatic hypotheses which are truths for faith; but it must recognize by unmistakable characters the one true religion, that is to say, that which alone merits the name of religion in that it unites all the characters which agree with that great and universal aspiration of the human soul.

One only thing, which is to all most evidently divine, is manifested in the world.

It is charity.

The work of true religion should be to produce, to preserve, and to spread abroad the Spirit of Charity.

To arrive at this end she must herself possess all the characteristics of charity, in such a manner that one could define her satisfactorily, in naming her: 'Organic Charity'.

Now, what are the characteristics of Charity?

It is St. Paul who will tell us.

Charity is patient.

Patient like God, because it is eternal as He is. It suffers persecutions, and never persecutes others.

It is kindly and loving, calling to itself the little, and not repulsing the great.

It is without jealousy. Of whom, and of what, should it be jealous? Has it not that better part which shall not be taken away from it?

It is neither quarrelsome nor intriguing.

It is without pride, without ambition, without selfishness, without anger.

It never thinks evil, and never triumphs by injustice; for all its joy is comprehended in truth.

It endures everything, without ever tolerating evil.

It believes all; its faith is simple, submissive, hierarchical, and universal.

It sustains all, and never imposes burdens which it is not itself the first to carry.

Religion is patient—the religion of great thinkers and of martyrs.

It is benevolent like Christ and the Apostles, like Vincent de Paul, and like Fénelon.

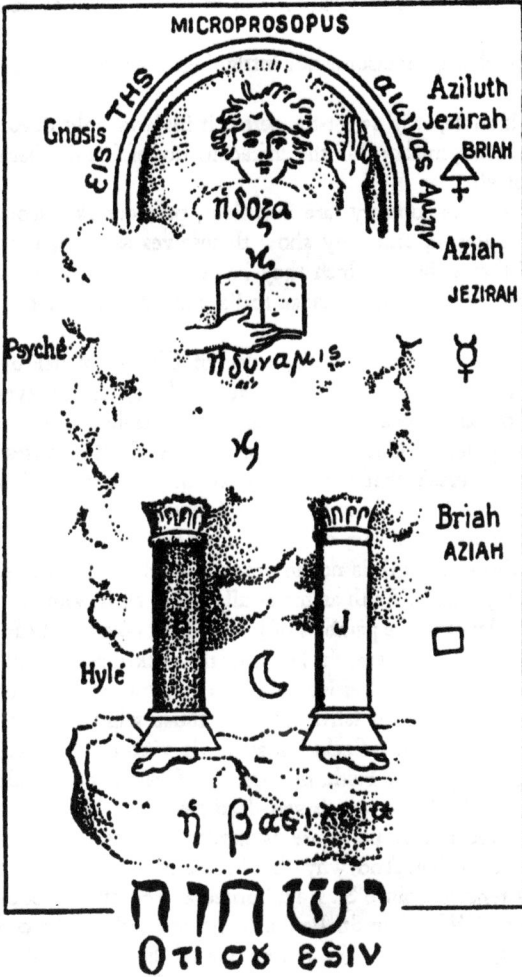

GREAT PENTACLE FROM THE VISION OF ST. JOHN

It envies not either the dignities or the goods of the earth. It is the religion of the fathers of the desert, of St. Francis, and of St. Bruno, of the Sisters of Charity, and of the Brothers of Saint-Jean-de-Dieu.

It is neither quarrelsome nor intriguing. It prays, does good, and waits.

It is humble, it is sweet-tempered, it inspires only devotion and sacrifice. It has, in short, all the characteristics of Charity because it is Charity itself.

Men, on the contrary, are impatient, persecutors, jealous, cruel, ambitious, unjust, and they show themselves as such, even in the name of that religion which they have succeeded in calumniating, but which they will never cause to lie. Men pass away, but truth is eternal.

Daughter of Charity, and creator of Charity in her own turn, true religion is essentially that which realizes; she believes in the miracles of faith, because she herself accomplishes them every day when she practises Charity. Now, a religion which practises Charity may flatter herself that she realizes all the dreams of divine love. Moreover, the faith of the hierarchical church transforms mysticism into realism by the efficacy of her sacraments. No more signs, no more figures whose strength is not in grace, and which do not really give what they promise! Faith animates all, makes all in some sort visible and palpable; even the parables of Jesus Christ take a body and a soul. They show, at Jerusalem, the house of the wicked rich man! The thin symbolisms of the primitive religions overturned by science, and deprived of the life of faith, resemble those whitened bones which covered the field that Ezekiel saw in his vision. The Spirit of the Saviour, the Spirit of Faith, the Spirit of Charity, has breathed upon this dust; and all that which was dead has taken life again so really that one recognizes no more yesterday's corpses in these living creatures of today. And why should one recognize them, since the world is renewed, since St. Paul burned at Ephesus the books of the hierophants? Was then St. Paul a barbarian, and was he committing a crime against science? No, but he burned the winding-sheets of the resuscitated that they might forget death. Why, then, do we today recall the qabalistic origins of dogma? Why do we join again the figures of the Bible to the allegories of Hermes? Is it to condemn St. Paul, is it to bring doubt to believers? No, indeed, for believers have no need of our book; they will not read it, and they will not wish to understand it. But we wish to show to the innumerable crowd of those who doubt, that faith is attached to the reason of all the centuries, to the science of all the sages. We wish to force human liberty to respect divine authority, reason to recognize the bases of faith, so that faith and authority, in their turn, may never again proscribe liberty and reason.

ARTICLE III

SOLUTION OF THE THIRD PROBLEM

THE RATIONALE OF THE MYSTERIES

FAITH being the aspiration to the unknown, the object of faith is absolutely and necessarily this one thing—Mystery.

In order to formulate its aspirations, faith is forced to borrow aspirations and images from the known.

But she specializes the employment of these forms, by placing them together in a manner which, in the known order of things, is impossible. Such is the profound reason of the apparent absurdity of symbolism.

Let us give an example:

If faith said that God was impersonal, one might conclude that God is only a word, or, at most, a thing.

If it is said that God was a person, one would represent to oneself the intelligent infinite, under the necessarily bounded form of an individual.

It says: 'God is one in three persons,' in order to express that one conceives in God both unity and multiplicity.

The formula of a mystery excludes necessarily the very intelligence of that formula, so far as it is borrowed from the world of known things; for, if one understood it, it would express the known and not the unknown.

It would then belong to science, and no longer to religion, that is to say, to faith.

The object of faith is a mathematical problem, those x escapes the procedures of our algebra.

Absolute mathematics prove only the necessity, and, in consequence, the existence of this unknown which we represent by the untranslatable x.

Now science progresses in vain; its progress is indefinite, but always relatively finite; it will never find in the language of the finite the complete expression of the infinite. Mystery is therefore eternal.

To bring into the logic of the known the terms of a profession of faith is to withdraw them from faith, which has for positive bases anti-logic, that is to say, the impossibility of logically explaining the unknown.

For the Jew, God is separate from humanity; He does not live in His creatures. He is infinite egoism.

For the Mussulman, God is a word before which one prostrates oneself, on the authority of Mohammed.

For the Christian, God has revealed Himself in humanity, proves Himself by Charity, and reigns by virtue of the order which constitutes the Hierarchy.

The Hierarchy is the guardian of dogma, for whose letter and spirit she alike demands respect. The sectarians who, in the name of their reason or, rather, of their individual unreason, have laid hands on dogma, have, in the very act, lost the Spirit of Charity; they have excommunicated themselves.

The Catholic, that is to say the universal, dogma merits that magnificent name by harmonizing in one all the religious aspirations of the world; with Moses and Mohammed, it affirms the unity of God; with Zoroaster, Hermes, and Plato, it recognizes in Him the infinite trinity of its own regeneration; it reconciles the living numbers of Pythagoras with the monadic Word of St. John;[1] so much, science and reason will agree. It is then in the eyes of reason and of science themselves the most perfect, that is to say the most complete, dogma which has ever been produced in the world. Let science and reason grant us so much; we shall ask nothing more of them.

'God exists; there is only one God, and He punishes those who do evil,' said Moses.

'God is everywhere; He is in us, and the good that we do to men we do it to God,' said Jesus.

'Fear' is the conclusion of the dogma of Moses.

'Love' is the conclusion of the dogma of Jesus.

The typical ideal of the life of God in humanity is incarnation.

Incarnation necessitates redemption, and operates it in the name

---

[1] The author had perhaps no space to continue with a demonstration that the Gospel legend itself is a macédoine of those of Bacchus, Adonis, Osiris, and a hundred others, and that the Mass, and Christian ceremonies generally, have similarly pagan sources.—A.C.

of the reversibility of solidarity,[1] or, in other words, of universal communion, the dogmatic principle of the Spirit of Charity.

To substitute human arbitrament for the legitimate despotism of the law, to put, in other words, tyranny in the place of authority, is the work of all Protestantism and of all democracies. What men call liberty is the sanction of illegitimate authority, or, rather, the fiction of power not sanctioned by authority.

John Calvin protested against the stakes of Rome, in order to give himself the right to burn Michael Servetus. Every people that liberates itself from a Charles I, or a Louis XVI, must undergo a Robespierre or a Cromwell, and there is a more or less absurd anti-pope behind all protestations against the legitimate papacy.

The divinity of Jesus Christ only exists in the Catholic Church, to which He transmits hierarchically His life and His divine powers. This divinity is sacerdotal and royal by virtue of communion; but outside of that communion, every affirmation of the divinity of Jesus Christ is idolatrous, because Jesus Christ could not be an isolated God.

The number of Protestants is of no importance to Catholic truth.

If all men were blind, would that be a reason for denying the existence of the sun?

Reason, in protesting against dogma, proves sufficiently that she has not invented it; but she is forced to admire the morality which results from that dogma. Now, if morality is a light, it follows that dogma must be a sun; light does not come from shadows.

Between the two abysses of polytheism, and an absurd and ignorant theism, there is only one possible medium: the mystery of the most Holy Trinity.

Between speculative theism, and anthropomorphism, there is only one possible medium: the mystery of incarnation.

Between immoral fatality, and Draconic responsibility, which would conclude the damnation of all beings, there is only one possible mean: the mystery of redemption.

The Trinity is Faith.

The Incarnation is Hope.

The Redemption is Charity.

The Trinity is the Hierarchy.

Incarnation is the Divine Authority of the Church.

[1] This and many similar phrases employed in the controversies of the period are today practically unintelligible. Lévi was at one time a kind of Socialist.—A.C.

Redemption is the unique, infallible, unfailing and Catholic priesthood.

The Catholic Church alone possesses an invariable dogma, and by its very constitution is incapable of corrupting morality; she does not make innovations, she explains. Thus, for example, the dogma of the immaculate conception is not new; it was contained in the theotokon of the Council of Ephesus, and the theotokon is a rigorous consequence of the Catholic dogma of the incarnation.

In the same way the Catholic Church makes no excommunications, she declares them; and she alone can declare them, because she alone is guardian of unity.

Outside the vessel of Peter, there is nothing but the abyss. Protestants are like people who have thrown themselves into the water in order to escape sea-sickness.

It is of Catholicity, such as it is constituted in the Roman Church, that one must say what Voltaire so boldly said of God: 'If it did not exist, it would be necessary to invent it.' But if a man had been capable of inventing the Spirit of Charity, he also would have invented God. Charity does not invent itself, it reveals itself by its works, and it is then that one can cry with the Saviour of the world: 'Blessed are the pure in heart, for they shall see God!'

To understand the Spirit of Charity is to understand all mysteries.

ARTICLE IV

SOLUTION OF THE FOURTH PROBLEM

RELIGION PROVED BY THE OBJECTIONS WHICH PEOPLE OPPOSE TO IT

THE objections which one may make against religion may be made either in the name of science, or in the name of reason, or in the name of faith.

Science cannot deny the facts of the existence of religion, of its establishment and its influence upon the events of history.

It is forbidden to it to touch dogma; dogma belongs wholly to faith.

Science ordinarily arms itself against religion with a series of facts which it is her duty to appreciate, which, in fact, she does appreciate thoroughly, but which she condemns still more energetically than science does.

In doing that, science admits that religion is right, and herself wrong; she lacks logic, manifests the disorder which every angry passion introduces into the spirit of man, and admits the need that it has of being ceaselessly redressed and directed by the Spirit of Charity.

Reason, on its side, examines dogma and finds it absurd.

But, if it were not so, reason would understand it; if reason understood it, it would no longer be the formula of the unknown.

It would be a mathematical demonstration of the infinite.

It would be the infinite finite, the unknown known, the immeasurable measured, the indicible named.

That is to say that dogma could only cease to be absurd in the eyes of reason to become, in the eyes of faith, science, reason and good sense in one, the most monstrous and the most impossible of all absurdities.

Remain the objections of dissent.

The Jews, our fathers in religion, reproach us with having attacked the unity of God, with having changed the immutable and eternal law, with adoring the creature instead of the Creator.

These heavy reproaches are founded on their perfectly false notion of Christianity.

Our God is the God of Moses, unique, immaterial, infinite God, sole object of worship, and ever the same.

Like the Jews, we believe Him to be present everywhere, but, as they ought to do, we believe Him living, thinking, and loving in humanity, and we adore Him in His works.

We have not changed His law, for the Jewish Decalogue is also the law of Christians.

The law is immutable because it is founded on the eternal principles of Nature; but the worship necessitated by the needs of man may change, and modify itself, parallel with the changes in men themselves.

This signifies that the worship itself is immutable, but modifies itself as language does.

Worship is a form of instruction; it is a language; one must translate it when nations no longer understand it.

We have translated, and not destroyed, the worship of Moses and of the prophets.

In adoring God in creation, we do not adore the creation itself.

In adoring God in Jesus Christ, it is God alone whom we adore, but God united to humanity.

In making humanity divine, Christianity has revealed the human divinity.

The God of the Jews was inhuman, because they did not understand Him in His works.

We are, then, more Israelite than the Israelites themselves. What they believe, we believe with them, and better than they do. They accuse us of having separated ourselves from them, and, on the contrary, it is they who wish to separate from us.

We wait for them, the heart and the arms wide open.

We are, as they are, the disciples of Moses.

Like them, we come from Egypt, and we detest its slavery. But we have entered into the Promised Land, and they obstinately abide and die in the desert.

Mohammedans are the bastards of Israel, or rather, they are his disinherited brothers, like Esau.

Their belief is illogical, for they admit that Jesus is a great prophet, and they treat Christians as infidels.

They recognize the Divine inspiration of Moses, yet they do not look upon the Jews as their brothers.

They believe blindly in their blind prophet, the fatalist Mohammed, the enemy of progress and of liberty.

Nevertheless, do not let us take away from Mohammed the glory of having proclaimed the unity of God among the idolatrous Arabs.

There are pure and sublime pages in the Qur'án.

In reading those pages, one may say with the children of Ishmael: 'There is no other God but God, and Mohammed is his prophet.'

There are three thrones in Heaven for the three prophets of the nations; but, at the end of time, Mohammed will be replaced by Elias.

The Mussulmans do not reproach the Christians; they insult them.

They call them infidels and 'giaours', that is to say, dogs. We have nothing to reply to them.

One must not refute the Turks and the Arabs; one must instruct and civilize them.

Remain dissident Christians, that is to say, those who, having broken the bond of unity, declare themselves strangers to the Charity of the Church.

Greek orthodoxy, that twin of the Roman Church which has not grown greater since its separation, which counts no longer in religion, which, since Photius, has not inspired a single eloquence, is a Church become entirely temporal, whose priesthood is no more than a function regulated by the imperial policy of the Tsar of All the Russias; a curious mummy of the primitive Church, still coloured and gilded with all its legends and all its rites, which its Popes no longer understand; the shadow of a living Church, but one which insisted on stopping when that Church moved on, and which is now no more than its bloated-out and headless silhouette.

Then, the Protestants, those eternal regulators of anarchy, who have broken down dogma, and are trying always to fill the void with reasonings, like the sieve of the Danaides; these weavers of religious fantasy, all of whose innovations are negative, who have formulated for their own use an unknown calling itself better known, mysteries better explained, a more defined infinite, a more restrained immensity, a more doubting faith, those who have quintessentialized the absurd, divided charity, and taken acts of anarchy for the principles of an entirely impossible hierarchy; those men who wish to realize salvation by faith alone, because charity escapes them, and who can no longer realize it, even upon the earth, for their pretended sacraments are no longer anything but allegorical mummeries; they no longer give grace; they no longer make God seen and touched; they are no longer, in a word, the signs of the almighty power of faith, but the compelled witnesses of the eternal impotence of doubt.

It is, then, against faith itself that the Reformation protested! Protestants were right only in their protest against the inconsiderate and persecuting zeal which wished to force consciences. They claimed the right to doubt, the right to have less religion than others, or even to have none at all; they have shed their blood for that sad privilege; they conquered it, they possess it; but they will not take away from us that of pitying them, and loving them. When the need to believe again takes them, when their heart revolts against the tyranny of a falsified reason, when they become tired of the empty abstractions of their arbitrary dogma, of the vague observances of their ineffective worship; when their communion without the real presence, their churches without divinity, and their morality without grace finally frighten them; when they are sick with the nostalgia of God—will they not rise up like the prodigal son, and come to throw themselves at the feet of the successor of Peter, saying: 'Father, we have sinned against Heaven and in thy sight, and we are no more worthy to be called thy sons, but count us among the humblest of thy servants'?

We will not speak of the criticism of Voltaire. That great mind was dominated by an ardent love of truth and justice, but he lacked that rectitude of heart which the intelligence of faith gives. Voltaire could not admit faith, because he did not know how to love. The spirit of charity did not reveal itself to that soul which had no tenderness, and he bitterly criticized the hearth of which he did not feel the warmth, and the lamp of which he did not see the light. If religion were such as he saw it, he would have been a thousand times right to attack it, and one would be obliged to fall on one's knees before the heroism of his courage. Voltaire would be the Messiah of good sense, the Hercules destructor of fanaticism. . . . But he laughed too much to understand Him who said: 'Happy are they who weep,' and the philosophy of laughter will never have anything in common with the religion of tears.

Voltaire parodied the Bible, dogma and worship; and then he mocked and insulted that parody.

Only those who recognize religion in Voltaire's parody can take offence at it. The Voltaireans are like the frogs in the fable who leap upon the log, and then make fun of royal majesty. They are at liberty to take the log for a king, they are at liberty to make once more that Roman caricature of which Tertullian once made mirth, that which represented the God of the Christians under the figure of a man with an ass's head. Christians will shrug their shoulders when they see this knavery, and pray God for the poor ignorants who imagine that they insult them.

M. the Count Joseph de Maistre, after having, in one of his most eloquent paradoxes, represented the hangman as a sacred being, and a permanent incarnation of divine justice upon earth, suggested that one should raise to the old man of Ferney a statue executed by the hangman. There is depth in this thought. Voltaire, in effect, also was, in the world, a being at the same time providential and fatal, endowed with insensibility for the accomplishment of his terrible functions. He was, in the domain of intelligence, a hangman, an exterminator armed by the justice of God Himself.

God sent Voltaire between the century of Bossuet and that of Napoleon in order to destroy everything that separates those two geniuses and to unite them in one alone.

He was the Samson of the spirit, always ready to shake the columns of the temple; but in order to make him turn in spite of himself the mill of religious progress, Providence made him blind of heart.

ARTICLE V

TO SEPARATE RELIGION FROM SUPERSTITION AND FANATICISM

SUPERSTITION, from the Latin word *superstes*, surviving, is the sign which survives the idea which it represents; it is the form preferred to the thing, the rite without reason, faith become insensate through isolating itself. It is in consequence the corpse of religion, the death of life, stupefaction substituted for inspiration.

Fanaticism is superstition become passionate, its name comes from the word *fanum*, which signifies 'temple', it is the temple put in the place of God, it is the human and temporal interest of the priest substituted for the honour of priesthood, the wretched passion of the man exploiting the faith of the believer.

In the fable of the ass loaded with relics, La Fontaine tells us that the animal thought that he was being adored; he did not tell us that certain people indeed thought that they were adoring the animal. These people were the superstitious.

If anyone had laughed at their stupidity, he would very likely have been assassinated, for from superstition to fanaticism is only one step.

Superstition is religion interpreted by stupidity; fanaticism is religion serving as a pretext to fury.

Those who intentionally and maliciously confound religion itself with superstition and fanaticism, borrow from stupidity its blind prejudices, and would borrow perhaps in the same way from fanaticism its injustices and angers.

Inquisitors or Septembrisors,[1] what matter names? The religion of Jesus Christ condemns, and has always condemned, assassins.

---

[1] Those who took part in the massacres of the Revolution of the 4th September 1792. —A.C.

RÉSUMÉ OF THE FIRST PART

In the Form of a Dialogue

FAITH, SCIENCE, REASON

SCIENCE: You will never make me believe in the existence of God.

FAITH: You have not the privilege of believing, but you will never prove to me that God does not exist.

SCIENCE: In order to prove it to you, I must first know what God is.

FAITH: You will never know it. If you knew it, you could teach it to me; and when I knew it, I should no longer believe it.

SCIENCE: Do you then believe without knowing what you believe?

FAITH: Oh, do not let us play with words! It is you who do not know what I believe, and I believe it precisely because you do not know it. Do you pretend to be infinite? Are you not stopped at every step by mystery? Mystery is for you an infinite ignorance which would reduce to nothing your finite knowledge, if I did not illumine it with my burning aspirations; and if, when you say: 'I no longer know,' I did not cry: 'As for me, I begin to believe.'

SCIENCE: But your aspirations and their object are not (and cannot be for me) anything but hypotheses.

FAITH: Doubtless, but they are certainties for me, since without those hypotheses I should be doubtful even about your certainties.

SCIENCE: But if you begin where I stop, you begin always too rashly and too soon. My progress bears witness that I am ever advancing.

FAITH: What does your progress matter, if I am always walking in front of you?

SCIENCE: You, walking? Dreamer of eternity, you have disdained earth too much; your feet are benumbed.

FAITH: I make my children carry me.

SCIENCE: They are the blind carrying the blind; beware of precipices!

FAITH: No, my children are by no means blind; on the contrary, they enjoy twofold sight: they see, by thine eyes, what thou canst show them upon earth, and they contemplate, by mine, what I show them in Heaven.

SCIENCE: What does Reason think of it?

REASON: I think, my dear teachers, that you illustrate a touching fable, that of the blind man and the paralytic. Science reproaches Faith with not knowing how to walk upon the earth, and Faith

74

says that Science sees nothing of her aspirations and of eternity in the sky. Instead of quarrelling, Science and Faith ought to unite; let Science carry Faith, and let Faith console Science by teaching her to hope and to love!

SCIENCE: It is a fine ideal, but Utopian. Faith will tell me absurdities. I prefer to walk without her.

FAITH: What do you call absurdities?

SCIENCE: I call absurdities propositions contrary to my demonstrations; as, for example, that three make one, that a God has become man, that is to say, that the Infinite has made itself finite, that the Eternal died, that God punished his innocent Son for the sin of guilty men. . . .

FAITH: Say no more about it. As enunciated by you, these propositions are in fact absurdities. Do you know what is the number of God, you who do not know God? Can you reason about the operations of the unknown? Can you understand the mysteries of Charity? I must always be absurd for you; for, if you understood them, my affirmations would be absorbed by your theorems; I should be you, and you would be me; or, to put it better, I should no longer exist, and Reason, in the presence of the infinite, would halt, blinded for evermore by your doubts, which are as infinite as space.

SCIENCE: At least, you should never usurp my authority, or give me the lie in my own domains.

FAITH: I have never done so, and I could never do so.

SCIENCE: So! You have never believed, for example, that a virgin could become a mother, without ceasing to be a virgin, in the physical, natural, positive order of things, in spite of all the laws of Nature; you do not affirm that a piece of bread is not only a God, but a real human body with its bones and its veins, its organs, its blood; such, in short, that you make of your children who eat this bread a little race of cannibals.

FAITH: There is not a single Christian who would not be revolted by what you have just said. That sufficiently proves that they do not understand my teachings in this positive and gross sense. The supernatural which I affirm is above Nature, and consequently could never oppose itself to her; the words of Faith are understood only by Faith; Science has only to repeat them to pervert their proper sense. I employ these words because I have no others; but since you find my remarks absurd, you should conclude from that that I give to the same words a meaning which escapes you. When the Saviour revealed the dogma of the real presence, did He not say: 'The flesh profiteth nothing, but my words are spirit and life.' I do not give you the mystery of the incarnation for an anatomical phenomenon, nor that of transubstantiation for a chemical manipulation. By what right do you exclaim 'Absurd!'? I do not reason about any of the things you know; by what right do you say that I talk unreasonably?

SCIENCE: I begin to understand you, or rather, I see that I shall never understand you. This being so, let us remain separate; I shall never have need of you.

FAITH: I am less proud, and I recognize that you may be useful to me perhaps. Perhaps, too, you will be very sad and very desperate without me, and I will not leave you unless Reason consents.

REASON: Be well 'ware of doing so! I am necessary to both of you. And I, what should I do without you? In order to be just, I need to know and to believe. But I must never confound what I know with what I believe. To know is to believe no longer; to believe is not to know as yet. The object of Science is the known; Faith does not occupy itself with that, and leaves it all to Science. The object of Faith is the unknown; Science may seek it, but not define it; she is then obliged, at least provisionally, to accept the definitions of Faith which it is impossible for her even to criticize. Only, if Science renounces Faith, she renounces hope and love, whose existence and necessity are as evident for Science as for Faith. Faith, as a psychological fact, pertains to the realm of Science; and Science, as the manifestation of the light of God within the human intelligence, pertains to the realm of Faith. Science and Faith must then admit each other, respect each other mutually, support each other, and bear each other aid in case of need, but without ever encroaching the one upon the other. The means of uniting them is—never to confound them. Never can there be contradiction between them, for although they use the same words, they do not speak the same language.

FAITH: Oh, well, Sister Science; what do you say about it?

SCIENCE: I say that we are separated by a deplorable misunderstanding, and that henceforward we shall be able to walk together. But to which of your different creeds do you wish to attach me? Shall I be Jewish, Catholic, Mohammedan, or Protestant?

FAITH: You will remain Science, and you will be universal.

SCIENCE: That is to say, Catholic, if I understand you correctly. But what should I think of the different religions?

FAITH: Judge them by their works. Seek true Charity, and when you have found her, ask her to which religion she belongs.

SCIENCE: It is certainly not to that of the Inquisition, and of the authors of the Massacre of St. Bartholomew.

FAITH: It is to that of St. John the Almoner, of St. François de Sales, of St. Vincent de Paul, of Fénelon, and so many more.

SCIENCE: Admit that if religion has produced much good, she has also done much evil.

FAITH: When one kills in the name of the God who said: 'Thou shalt not kill',[1] when one persecutes in the name of Him who commands us to forgive our enemies, when one propagates darkness in the name of Him who tells us not to hide the light under a bushel, is it just to attribute the crime to the very law which condemns it? Say, if you wish to be just, that in spite of religion, much evil has been done upon earth. But also, to how many virtues has it not given birth? How many are the devotions, how many the sacrifices, of which we do not know! Have you counted those noble hearts, both men and women, who renounced all joys to enter the service of all sorrows? Those souls devoted to labour and to prayer, who have strewn their pathways with good deeds? Who founded asylums for orphans and old men, hospitals for the sick, retreats for the repentant? These institutions, as glorious as they are modest, are the real works with which the annals of the Church are filled; religious wars and the persecution of heretics belong to the politics of savage centuries. The heretics, moreover, were themselves murderers. Have you forgotten the burning of Michael Servetus and the massacre of our priests, renewed, still in the name of humanity and reason, by the revolutionaries who hated the Inquisition and the Massacre of St. Bartholomew? Men are always cruel, it is true, but only when they forget the religion whose watchwords are blessing and pardon.

SCIENCE: O Faith! Pardon me, then, if I cannot believe; but I know now why you believe. I respect your hopes, and share your desires. But I must find by seeking; and in order to seek, I must doubt.

REASON: Work, then, and seek, O Science, but respect the oracles of Faith! When your doubt leaves a gap in universal enlightenment, allow Faith to fill it! Walk distinguished the one from the other, but leaning the one upon the other, and you will never go astray.

[1] And habitually commanded the rape of virgins and the massacre of children. I Sam. xv. 3, etc.—A.C.

PART II

Philosophical Mysteries

PRELIMINARY CONSIDERATIONS

IT HAS been said that beauty is the splendour of truth.

Now moral beauty is goodness. It is beautiful to be good.

To be intelligently good, one must be just.

To be just, one must act reasonably.

To act reasonably, one must have the knowledge of reality.

To have the knowledge of reality, one must have consciousness of truth.

To have consciousness of truth, one must have an exact notion of being.

Being, truth, reason, and justice are the common objects of the researches of science, and of the aspirations of faith. The conceptions, whether real or hypothetical, of a supreme power, transform justice into Providence; and the notion of divinity, from this point of view, becomes accessible to science herself.

Science studies Being in its partial manifestation; faith supposes it, or rather admits it *a priori* as a whole.

Science seeks the truth in everything; faith refers everything to an universal and absolute truth.

Science records realities in detail; faith explains them by totalized reality to which science cannot bear witness, but which the very existence of the details seems to force her to recognize and to admit.

Science submits the reasons of persons and things to the universal mathematical reason; faith seeks, or rather supposes, an intelligent and absolute reason for (and above) mathematics themselves.

Science demonstrates justice by justness; faith gives an absolute justness to justice, in subordinating it to Providence.

One sees here all that faith borrows from science, and all that science, in its turn, owes to faith.

Without faith, science is circumscribed by an absolute doubt, and finds itself eternally penned within the risky empiricism of a reasoning scepticism; without science, faith constructs its hypotheses at random, and can only blindly prejudge the causes of the effects of which she is ignorant.

The great chain which reunites science and faith is analogy.

Science is obliged to respect a belief whose hypotheses are analogous to demonstrated truths. Faith, which attributes everything to God, is obliged to admit science as being a natural revelation which, by the partial manifestation of the laws of eternal reason, gives a scale of proportion to all the aspirations and to all the excursions of the soul into the domain of the unknown.

It is, then, faith alone that can give a solution to the mysteries of science; and in return, it is science alone that demonstrates the necessity of the mysteries of faith.

Outside the union and the concourse of these two living forces of the intelligence, there is for science nothing but scepticism and despair, for faith nothing but rashness and fanaticism.

If faith insults science, she blasphemes; if science misunderstands faith, she abdicates.

Now let us hear them speak in harmony!

'Being is everywhere,' says science, 'it is multiple and variable in its forms, unique in its essence, and immutable in its laws. The relative demonstrates the existence of the absolute. Intelligence exists in being. Intelligence animates and modifies matter.'

'Intelligence is everywhere,' says faith; 'Life is nowhere fatal because it is ruled. This rule is the expression of supreme Wisdom. The absolute in intelligence, the supreme regulator of forms, the living ideal of spirits, is God.'

'In its identity with the ideal, being is truth,' says science.

'In its identity with the ideal, truth is God,' replies faith.

'In its identity with my demonstrations, being is reality,' says science.

'In its identity with my legitimate aspirations, reality is my dogma,' says faith.

'In its identity with the Word, being is reason,' says science.

'In its identity with the Spirit of Charity, the highest reason is my obedience,' says faith.

'In its identity with the motive of reasonable acts, being is justice,' says science.

'In its identity with the principle of Charity, justice is Providence,' replies faith.

Sublime harmony of all certainties with all hopes, of the absolute in intelligence with the absolute in love! The Holy Spirit, the Spirit of Charity, should then conciliate all, and transform all into His own light. Is it not the Spirit of Intelligence, the Spirit of Science, the Spirit of Counsel, the Spirit of Force? 'He must come,' says the Catholic liturgy, 'and it will be, as it were, a new creation; and He will change the face of the earth.'

'To laugh at philosophy is already to philosophize,' said Pascal, referring to that sceptical and incredulous philosophy which does not recognize faith. And if there existed a faith which trampled science underfoot, we should not say that to laugh at such a faith would be a true act of religion, for religion, which is all Charity, does not tolerate mockery; but one would be right in blaming this love for ignorance, and in saying to this rash faith, 'Since you slight your sister, you are not the daughter of God!'

Truth, reality, reason, justice, Providence, these are the five rays of the flamboyant star in the centre of which science will write the word 'being'—to which faith will add the ineffable name of God.

SOLUTION OF THE PHILOSOPHICAL PROBLEMS

First Series

Q. What is truth?
A. Idea identical with being.
Q. What is reality?
A. Knowledge identical with being.
Q. What is reason?
A. The Word identical with being.
Q. What is justice?
A. The motive of acts identical with being.
Q. What is the absolute?
A. Being.
Q. Can one conceive anything superior to being?
A. No; but one conceives in being itself something supereminent and transcendental.
Q. What is that?
A. The supreme reason of being.

Q. Do you know it, and can you define it?
A. Faith alone affirms it, and names it God.
Q. Is there anything above truth?
A. Above known truth, there is unknown truth.
Q. How can one construct reasonable hypotheses with regard to this truth?
A. By analogy and proportion.
Q. How can one define it?
A. By the symbols of faith.
Q. Can one say of reality the same thing as of truth?
A. Exactly the same thing.
Q. Is there anything above reason?
A. Above finite reason, there is infinite reason.
Q. What is infinite reason?
A. It is that supreme reason of being that faith calls God.
Q. Is there anything above justice?
A. Yes; according to faith, there is the Providence of God, and the sacrifice of man.
Q. What is this sacrifice?
A. It is the willing and spontaneous surrender of right.
Q. Is this sacrifice reasonable?
A. No; it is a kind of folly greater than reason, for reason is forced to admire it.
Q. How does one call a man who acts according to truth, reality, reason and justice?
A. A moral man.
Q. And if he sacrifices his interests to justice?
A. A man of honour.
Q. And if in order to imitate the grandeur and goodness of Providence he does more than his duty, and sacrifices his right to the good of others?
A. A hero.
Q. What is the principle of true heroism?
A. Faith.
Q. What is its support?
A. Hope.
Q. And its rule?
A. Charity.
Q. What is the Good?
A. Order.

Q. What is the Evil?

A. Disorder.

Q. What is permissible pleasure?

A. Enjoyment of order.

Q. What is forbidden pleasure?

A. Enjoyment of disorder.

Q. What are the consequences of each?

A. Moral life and moral death.

Q. Has then hell, with all its horrors, its justification in religious dogma?

A. Yes; it is a rigorous consequence of a principle.

Q. What is this principle?

A. Liberty.

Q. What is liberty?

A. The right to do one's duty, with the possibility of not doing it.

Q. What is failing in one's duty?

A. It involves the loss of one's right. Now, right being eternal, to lose it is to suffer an eternal loss.

Q. Can one repair a fault?

A. Yes; by expiation.

Q. What is expiation?

A. Working overtime. Thus, because I was lazy yesterday, I had to do a double task today.

Q. What are we to think of those who impose on themselves voluntary sufferings?

A. If they do so in order to overcome the brutal fascination of pleasure, they are wise; if to suffer instead of others, they are generous; but if they do it without discretion and without measure, they are imprudent.

Q. Thus, in the eyes of true philosophy, religion is wise in all that it ordains?

A. You see that it is so.

Q. But if, after all, we were deceived in our eternal hopes?

A. Faith does not admit that doubt. But philosophy herself should reply that all the pleasures of the earth are not worth one day of wisdom, and that all the triumphs of ambition are not worth a single minute of heroism and of Charity.

Q. What is man?

A. Man is an intelligent and corporeal being made in the image of God and of the world, one in essence, triple in substance, mortal and immortal.

Q. You say, 'triple in substance'. Has man, then, two souls or two bodies?

A. No; there is in him a spiritual soul, a material body, and a plastic medium.

Q. What is the substance of this medium?

A. Light, partially volatile, and partially fixed.

Q. What is the volatile part of this light ?

A. Magnetic fluid.

Q. And the fixed part?

A. The fluidic or fragrant body.

Q. Is the existence of this body demonstrated?

A. Yes; by the most curious and the most conclusive experiences. We shall speak of them in the third part of this work.

Q. Are these experiences articles of faith?

A. No, they pertain to science.

Q. But will science preoccupy herself with it?

A. She already preoccupies herself with it. We have written this book, and you are reading it.

Q. Give us some notions of this plastic medium.

A. It is formed of astral or terrestrial light, and transmits the double magnetization of it to the human body. The soul, by acting on this light through its volitions, can dissolve it or coagulate it, project it or withdraw it. It is the mirror of the imagination and of dreams. It reacts upon the nervous system, and thus produces the movements of the body. This light can dilate itself indefinitely, and communicate its reflections at considerable distances; it magnetizes the bodies submitted to the action of man, and can, by concentrating itself, again draw them to him. It can take all the forms evoked by thought, and, in the transitory coagulations of its radiant particles, appear to the eyes; it can even offer a sort of resistance to the touch. But these manifestations and uses of the plastic medium being abnormal, the luminous instrument of precision cannot produce them without being strained, and there is danger of either habitual hallucination, or of insanity.

Q. What is animal magnetism?

A. The action of one plastic medium upon another, in order to dissolve or coagulate it. By augmenting the elasticity of the vital light and its force of projection, one sends it forth as far as one will, and withdraws it completely loaded with images; but this operation must be favoured by the slumber of the subject, which one produces by coagulating still further the fixed part of his medium.

Q. Is magnetism contrary to morality and religion?

A. Yes, when one abuses it.

Q. In what does the abuse of it consist?

A. In employing it in a disordered manner, or for a disordered object.

Q. What is a disordered magnetism?

A. An unwholesome fluidic emission, made with a bad intention; for example, to know the secrets of others, or to arrive at unworthy ends.

Q. What is the result of it?

A. It puts out of order the fluidic instrument of precision, both in the case of the magnetizer and of the magnetized. To this cause one must attribute the immoralities and the follies with which a great number of those who occupy themselves with magnetism are reproached.

Q. What conditions are required in order to magnetize properly?

A. Health of spirit and body; right intention, and discreet practice.

Q. What advantageous results can one obtain by discreet magnetism?

A. The cure of nervous diseases, the analysis of presentiments, the re-establishment of fluidic harmonies, and the rediscovery of certain secrets of Nature.

Q. Explain that to us in a more complete manner.

A. We shall do so in the third part of this work, which will treat specially of the mysteries of Nature.

PART III

The Mysteries of Nature

THE GREAT MAGICAL AGENT

ARCHÉE

AZOTH

HYLE

THE TENTH KEY OF THE TAROT

WE HAVE spoken of a substance extended in the infinite.

That substance is one which is Heaven and earth; that is to say, according to its degrees of polarization, subtle or fixed.

This substance is what Hermes Trismegistus calls the great *Telesma*. When it produces splendour, it is called Light.

It is this substance which God creates before everything else, when He says, 'Let there be light.'

It is at once substance and movement.

It is fluid, and a perpetual vibration.

Its inherent force which sets it in motion is called *magnetism*.

In the infinite, this unique substance is the ether, or the etheric light.

In the stars which it magnetizes, it becomes astral light.

In organized beings, light, or magnetic fluid.

In man it forms the *astral body*, or the *plastic medium*.

The will of intelligent beings acts directly on this light, and by means of it on all that part of Nature which is submitted to the modifications of intelligence.

This light is the common mirror of all thoughts and all forms; it preserves the images of everything that has been, the reflections of past worlds, and, by analogy, the sketches of worlds to come. It is the instrument of thaumaturgy and divination, as remains for us to explain in the third and last part of this work.

FIRST BOOK

MAGNETIC MYSTERIES

CHAPTER I

The Key of Mesmerism

MESMER rediscovered the secret science of Nature; he did not invent it.

The first unique and elementary substance whose existence he proclaims in his aphorisms, was known by Hermes and Pythagoras.

Synesius, who sings it in his hymns, had found it revealed in the Platonistic records of the School of Alexandria:

Μία παγὰ, μία ῥίζα
Τριφάης ἔλαμπε μορφᾳ
. . . .
Περὶ γὰρ σπάρεισα πνοιὰ
Χθονὸς ἐζώωσε μοίρας
Πολυδαιδάλοισι μόραις

A single source, a single root of light, jets out and spreads itself into three branches of splendour. A breath blows round the earth, and vivifies in innumerable forms all parts of animated substance.'—

Hymn II—*Synesius*

Mesmer saw in elementary matter a substance indifferent to movement as to rest. Submitted to movement, it is volatile; fallen back into rest, it is fixed; and he did not understand that movement is inherent in the first substance; that it results, not from its indifference, but from its aptitude, combined with a movement and a rest which are equilibrated the one by the other; that absolute rest is nowhere in universal living matter, but that the fixed attracts the volatile in order to fix it; while the volatile attacks the fixed in order to volatilize it. That the supposed rest of particles apparently fixed, is nothing but a more desperate struggle and a greater tension of their fluidic forces, which by neutralizing each other make themselves immobile. It is thus that, as Hermes says, that which is above is like that which is below; the same force which expands steam, contracts and hardens the icicle; everything obeys the laws of life which are inherent in the original substance; this substance attracts and repels, coagulates itself and dissolves itself, with a constant harmony; it is double; it is androgynous; it embraces itself, and fertilizes itself, it struggles, triumphs, destroys, renews; but never abandons itself to inertia, because inertia, for it, would be death.

It is this original substance to which the hieratic recital of Genesis refers when the word of the Elohim creates light by commanding it to exist.

The Elohim said: 'Let there be light!' and there was light.

This light, whose Hebrew name is אור, *aour*, is the fluidic and living gold of the hermetic philosophy. Its positive principle is their sulphur; its negative principle, their mercury; and its equilibrated principles form what they call their salt.

One must then, in place of the sixth aphorism of Mesmer which reads thus: 'Matter is indifferent as to whether it is in movement or at rest,' establish this proposition: 'The universal matter is compelled to movement by its double magnetization, and its fate is to seek equilibrium.'

Whence one may deduce these corollaries:

Regularity and variety in movement result from the different combinations of equilibrium.

A point equilibrated on all sides remains at rest, for the very reason that it is endowed with motion.

A fluid consists of rapidly moving matter, always stirred by the variation of the balancing forces.

A solid is the same matter in slow movement, or at apparent rest because it is more or less solidly balanced.

There is no solid body which would not immediately be pulverized, vanish in smoke, and become invisible if the equilibrium of its molecules were to cease suddenly.

There is no fluid which would not instantly become harder than the diamond, if one could equilibrate its constituent molecules.

To direct the magnetic forces is then to destroy or create forms; to produce to all appearance, or to destroy bodies; it is to exercise the almighty power of Nature.

Our plastic medium is a magnet which attracts or repels the astral light under the pressure of the will. It is a luminous body which reproduces with the greatest ease forms corresponding to ideas.

It is the mirror of the imagination. This body is nourished by astral light just as the organic body is nourished by the products of the earth. During slumber, it absorbs the astral light by immersion, and during waking, by a kind of somewhat slow respiration. When the phenomena of natural somnambulism are produced, the plastic medium is surcharged with ill-digested nourishment. The will, although bound by the torpor of slumber, repels instinctively the medium towards the organs in order to disengage it, and a reaction, of mechanical nature, takes place, which with the movement of the body equilibrates the light of the medium. It is for that reason that it is so dangerous to wake somnambulists suddenly, for the gorged medium may then withdraw itself suddenly towards the common reservoir, and abandon the organs altogether; these are then separated from the soul, and death is the result.

The state of somnambulism, whether natural or artificial, is then extremely dangerous, because in uniting the phenomena of the waking state and the state of slumber, it constitutes a sort of straddle between two worlds. The soul moves the springs of the particular life while bathing itself in the universal life, and experiences an inexpressible sense of well-being; it will then willingly let go the nervous branches which hold it suspended above the current. In ecstasies of every kind the situation is the same. If the will plunges into it with a passionate effort, or even abandons itself entirely to it, the subject may become insane or paralysed, or even die.

Hallucinations and visions result from wounds inflicted on the plastic medium, and from its local paralysis. Sometimes it ceases to give forth rays, and substitutes images condensed somehow or other to realities shown by the light; sometimes it radiates with too much force, and condenses itself outside and around some chance and irregulated nucleus, as blood does in some bodily growths. Then the chimeras of our brain take on a body, and seem to take on a soul; we appear to ourselves radiant or deformed according to the image of the ideal of our desires, or our fears.

Hallucinations, being the dreams of waking persons, always imply a state analogous to somnambulism. But in a contrary sense; somnambulism is slumber borrowing its phenomena from waking; hallucination is waking still partially subjected to the astral intoxication of slumber.

Our fluidic bodies attract and repulse each other following laws similar to those of electricity. It is this which produces instinctive sympathies and antipathies. They thus equilibrate each other, and for this reason hallucinations are often contagious; abnormal projections change the luminous currents; the perturbation caused by a sick person wins over to itself the more sensitive natures; a circle of illusions is established, and a whole crowd of people is easily dragged away thereby. Such is the history of strange apparitions and popular prodigies. Thus are explained the miracles of the American mediums and the hysterics of table-turners, who reproduce in our own times the ecstasies of whirling dervishes. The sorcerers of Lapland with their magic drums, and the conjurer medicine-men of savages, arrive at similar results by similar proceedings; their gods or their devils have nothing to do with it.

Madmen and idiots are more sensitive to magnetism than people of sound mind; it should be easy to understand the reason of that:

very little is required to turn completely the head of a drunken man, and one more easily acquires a disease when all the organs are predisposed to submit to its impressions, and manifest its disorders.

Fluidic maladies have their fatal crises. Every abnormal tension of the nervous apparatus ends in the contrary tension, according to the necessary laws of equilibrium. An exaggerated love changes to aversion, and every exalted hate comes very near to love; the reaction happens suddenly with the flame and violence of the thunderbolt. Ignorance then laments it or exclaims against it; science resigns itself, and remains silent.

There are two loves, that of the heart, and that of the head: the love of the heart never excites itself, it gathers itself together, and grows slowly by the path of ordeal and sacrifice; purely nervous and passionate, cerebral love lives only on enthusiasm, dashes itself against all duties, treats the beloved object as a prize of conquest, is selfish, exacting, restless, tyrannical, and is fated to drag after it either suicide as the final catastrophe, or adultery as a remedy. These phenomena are constant like nature, inexorable as fatality.

A young artist full of courage, with her future all before her, had a husband, an honest man, a seeker after knowledge, a poet, whose only fault was an excess of love for her; she outraged him and left him, and has continued to hate him ever since. Yet she, too, is a decent woman; the pitiless world, however, judges and condemns her. And yet, this was not her crime. Her fault, if one may be permitted to reproach her with one, was that, at first, she madly and passionately loved her husband.

'But,' you will say, 'is not the human soul, then, free?' No, it is no longer free when it has abandoned itself to the giddiness caused by passion. It is only wisdom which is free; disordered passions are the kingdom of folly, and folly is fatality.

What we have said of love may equally well be said of religion, which is the most powerful, but also the most intoxicating, of all loves. Religious passion has also its excesses and its fatal reactions. One may have ecstasies and stigmata like St. Francis of Assisi, and fall afterwards into abysses of debauch and impiety.

Passionate natures are highly charged magnets; they attract or repel with violence.

It is possible to magnetize in two ways: first, in acting by will upon the plastic medium of another person, whose will and whose acts are, in consequence, subordinated to that action.

Secondly, in acting through the will of another, either by intimidation, or by persuasion, so that the influenced will modifies at our pleasure the plastic medium and the acts of that person.

One magnetizes by radiation, by contact, by look, or by word.

The vibrations of the voice modify the movement of the astral light, and are a powerful vehicle of magnetism.

The warm breath magnetizes positively, and the cold breath negatively.

A warm and prolonged insufflation upon the spinal column at the base of the cerebellum may occasion erotic phenomena.

If one puts the right hand upon the head and the left hand under the feet of a person completely enveloped with wool or silk, one causes the magnetic spark to pass completely through the body, and one may thus occasion a nervous revolution in his organism with the rapidity of lightning.

Magnetic passes only serve to direct the will of the magnetizer in confirming it by acts. They are signs and nothing more. The act of the will is expressed and not operated by these signs.

Powdered charcoal absorbs and retains the astral light. This explains the magic mirror of Dupotet.

Figures traced in charcoal appear luminous to a magnetized person, and take, for him, following the direction indicated by the will of the magnetizer, the most gracious or the most terrifying forms.

The astral light, or rather the vital light, of the plastic medium, absorbed by the charcoal, becomes wholly negative; for this reason animals which are tormented by electricity, as for example, cats, love to roll themselves upon coal. One day, medicine will make use of this property, and nervous persons will find great relief from it.

CHAPTER II

Life and Death—Sleep and Waking

SLEEP is an incomplete death; death is a complete sleep.

Nature subjects us to sleep in order to accustom us to the idea of death, and warns us by dreams of the persistence of another life.

The astral light into which sleep plunges us is like an ocean in which innumerable images are afloat, flotsam of wrecked existences, mirages and reflections of those which pass, presentiments of those which are about to be.

Our nervous disposition attracts to us those images which correspond to our agitation, to the nature of our fatigue, just as a magnet, moved among particles of various metals, would attract to itself and choose particularly the iron filings.

Dreams reveal to us the sickness or the health, the calm or the disturbance, of our plastic medium, and consequently, also that of our nervous apparatus.

They formulate our presentiments by the analogy which the images bear to them.

For all ideas have a double significance for us, relating to our double life.

There exists a language of sleep; in the waking stage it is impossible to understand it, or even to order its words.

The language of slumber is that of nature, hieroglyphic in its character, and rhythmical in its sounds.

Slumber may be either giddy or lucid.

Madness is a permanent state of vertiginous somnambulism.

A violent disturbance may wake madmen to sense, or kill them.

Hallucinations, when they obtain the adhesion of the intelligence, are transitory attacks of madness.

Every mental fatigue provokes slumber; but if the fatigue is accompanied by nervous irritation, the slumber may be incomplete, and take on the character of somnambulism.

One sometimes goes to sleep without knowing it in the midst of real life; and then, instead of thinking, one dreams.

How is it that we remember things which have never happened to us? Because we dreamt them when wide awake.

This phenomenon of involuntary and unperceived sleep when it suddenly traverses real life, often happens to those who over-excite their nervous organism by excesses either of work, vigil, drink, or erethism.

Monomaniacs are asleep when they perform unreasonable acts. They no longer remember anything on waking.

When Papavoine was arrested by the police, he calmly said to them these remarkable words: *You are taking the other for me.*

It was the somnambulist who was still speaking.

Edgar Poe, that unhappy man of genius who used to intoxicate himself, has terribly described the somnambulism of monomaniacs.

Sometimes it is an assassin who hears, and who thinks that everybody hears, through the walls of the tomb, the beating of his victim's heart; sometimes it is a poisoner who, by dint of saying to himself: 'I am safe, provided I do not go and denounce myself,' ends by dreaming aloud that he is denouncing himself, and in fact does so. Edgar Poe himself invented neither the persons nor the facts of these strange novels; he dreamt them waking, and that is why he clothed them so well with all the colours of a shocking reality.

Dr. Brière de Boismont in his remarkable work on *Hallucinations*, tells the story of an Englishman otherwise quite sane, who thought that he had met a stranger and made his acquaintance, who took him to lunch at his tavern, and then having asked him to visit St. Paul's in his company, had tried to throw him from the top of the tower which they had climbed together.

From that moment the Englishman was obsessed by this stranger, whom he alone could see, and whom he always met when he was alone, and had dined well.

Precipices attract; drunkenness calls to drunkenness; madness has invincible charms for madness. When a man succumbs to sleep, he holds in horror everything which might wake him. It is the same with the hallucinated, with statical somnambulists, maniacs, epileptics, and all those who abandon themselves to the delirium of a passion. They have heard the fatal music, they have entered into the dance of death; and they feel themselves dragged away into the whirl of vertigo. You speak to them, they no more hear you; you warn them, they no longer understand you, but your voice annoys them; they are asleep with the sleep of death.

Death is a current which carries you away, a whirlpool which draws you down, but from the bottom of which the least movement may make you climb again. The force of repulsion being equal to that of attraction, at the very moment of expiring, one often attaches oneself again violently to life. Often also, by the same law of equilibrium, one passes from sleep to death through complaisance for sleep.

A shallop sways upon the shores of the lake. The child enters the water, which, shining with a thousand reflections, dances around him and calls him; the chain which retains the boat stretches and seems to wish to break itself; then a marvellous bird shoots out from the bank, and skims, singing, upon the joyous waves; the child wishes to follow it, he puts his hand upon the chain, he detaches the ring.

Antiquity divined the mystery of the attraction of death, and represented it in the fable of Hylas. Weary with a long voyage, Hylas has arrived in a flowered, enamelled isle; he approaches a fountain to draw water; a gracious mirage smiles at him; he sees a nymph stretch out her arms to him, his own lose nerve, and cannot draw back the heavy jar; the fresh fragrance of the spring puts him to sleep; the perfumes of the bank intoxicate him. There he is, bent over the water like a narcissus whose stalk has been broken by a child at play; the full jar falls to the bottom, and Hylas follows it; he dies, dreaming that nymphs caress him, and no longer hears the voice of Hercules recalling him to the labours of life; Hercules, who runs wildly everywhere, crying: 'Hylas! Hylas!'

Another fable, not less touching, which steps forth from the shadows of the Orphic initiation, is that of Eurydice recalled to life by the miracles of harmony and love, of Eurydice, that sensitive broken on the very day of her marriage, who takes refuge in the tomb, trembling with modesty. Soon she hears the lyre of Orpheus, and slowly climbs again towards the light; the terrible divinities of Erebus dare not bar her passage. She follows the poet, or rather the poetry which she adores. . . . But, woe to the lover if he changes the magnetic current and pursues in his turn, with a single look, her whom he should only attract! The sacred love, the virginal love, the love which is stronger than the tomb, seeks only devotion, and flies in terror before the egoism of desire. Orpheus knows it; but, for an instant, he forgets it. Eurydice, in her white bridal dress, lies upon the marriage bed; he wears the vestments of a Grand Hierophant, he stands upright, his lyre in his hand, his head crowned with the sacred laurel, his eyes turned towards the East, and he sings. He sings of the luminous arrows of love that traverse the shadows of old Chaos, the waves of soft, clear light, flowing from the black teats of the mother of the gods, from which hang the two children, Eros and Anteros. He sings the song of Adonis returning to life in answer to the complaint of Venus, reviving like a flower under the shining dew of her tears; the song of Castor and Pollux, whom death could not divide, and who love alternately in hell and upon earth. . . . Then he calls softly Eurydice, his dear Eurydice, his so much loved Eurydice:

> Ah! miseram Eurydicen animâ fugiente vocabat,
> Eurydicen! toto referebant flumine ripae.

While he sings, that pallid statue of the sculptor death takes on the colour of the first tint of life, its white lips begin to redden like the dawn... Orpheus sees her, he trembles, he stammers, the hymn almost dies upon his lips, but she pales anew; then the Grand Hierophant tears from his lyre sublime heartrending songs, he looks no more save upon Heaven, he weeps, he prays, and Eurydice opens her eyes ... Unhappy one, do not look at her! sing! sing! do not scare away the butterfly of Psyche, which is about to alight on this flower! But the insensate man has seen the look of the woman whom he has raised from the dead, the Grand Hierophant gives place to the lover, his lyre falls from his hands, he looks upon Eurydice, he darts towards her ... he clasps her in his arms, he finds her frozen still, her eyes are closed again, her lips are paler and colder than ever, the sensitive soul has trembled, the frail cord is broken anew—and for ever. . . . Eurydice is dead, and the hymns of Orpheus can no longer recall her to life!

In our *Transcendental Magic*, we had the temerity to say that the resurrection of the dead is not an impossible phenomenon even on the physical plane; and in saying that, we have not denied or in any way contradicted the fatal law of death. A death which can discontinue is only lethargy and slumber; but it is by lethargy and slumber that death always begins. The state of profound peace which succeeds the agitations of life carries away the relaxed and sleeping soul; one cannot make it return, and force it to plunge anew into life, except by exciting violently all its affections and all its desires. When Jesus, the Saviour of the world, was upon the earth, the earth was more beautiful and more desirable than Heaven; and yet it was necessary for Jesus to cry aloud and apply a shock in order to awaken Jairus's daughter. It was by dint of shudderings and tears that He called back his friend Lazarus from the tomb, so difficult is it to interrupt a tired soul who is sleeping his beauty-sleep!

At the same time, the countenance of death has not the same serenity for every soul that contemplates it. When one has missed the goal of life, when one carries away with one frenzied greeds or unas-suaged hates, eternity appears to the ignorant or guilty soul with such a formidable proportion of sorrows, that it sometimes tries to fling itself back into mortal life. How many souls, urged by the nightmare of hell, have taken refuge in their frozen bodies, their bodies already covered with funereal marble! Men have found skeletons turned over, con-vulsed, twisted, and they have said, 'Here are men who have been buried alive.' Often this was not the case. These may always be waifs of death, men raised from the tomb, who, before they could abandon themselves altogether to the anguish of the threshold of eternity, were obliged to make a second attempt.

A celebrated magnetist, Baron Dupotet, teaches in his secret book on *Magic* that one can kill by magic as by electricity. There is nothing strange in this revelation for anyone who is well acquainted with the analogies of Nature. It is certain that in diluting beyond measure, or in coagulating suddenly, the plastic medium of a subject, it is possible to loose the body from the soul. It is sometimes sufficient to arouse a violent anger, or an overmastering fear in anyone, to kill him suddenly.

The habitual use of magnetism usually puts the subject who adandons himself to it at the mercy of the magnetizer. When communication is well established, and the magnetizer can produce at will slumber, insensibility, catalepsy, and so on, it will only require a little further effort to bring on death.

We have been told as an actual fact a story whose authenticity we will not altogether guarantee.

We are about to repeat it because it may be true.

Certain persons who doubted both religion and magnetism, of that incredulous class which is ready for all superstitions and all fanaticisms, had persuaded a poor girl to submit to their experiments for a fee. This girl was of an impressionable and nervous nature, fatigued moreover by the excesses of a life which had been more than irregular, while she was already disgusted with existence. They put her to sleep; bade her see; she weeps and struggles. They speak to her of God; she trembles in every limb.

'No,' said she, 'no; He frightens me; I will not look at Him.'

'Look at Him, I wish it.'

She opens her eyes, her pupils expand; she is terrifying.

'What do you see?'

'I should not know how to say it. . . . Oh for pity's sake awaken me!'

'No, look, and say what you see.'

'I see a black night in which whirl sparks of every colour around two great ever-rolling eyes. From these eyes leap rays whose spiral whorls fill space. . . . Oh, it hurts me! Wake me!'

'No, look.'

'Where do you wish me to look now?'

'Look into Paradise.'

'No, I cannot climb there; the great night pushes me back, I always fall back.'

'Very well then, look into hell.'

Here the sleep-waker became convulsively agitated.

'No, no!' she cried sobbing; 'I will not! I shall be giddy; I should fall! Oh, hold me back! Hold be back!'

'No, descend.'

'Where do you want me to descend?'

'Into hell.'

'But it is horrible! No! No! I will not go there!'

'Go there.'

'Mercy!'

'Go there. It is my will.'

The features of the sleep-waker become terrible to behold; her hair stands on end; her wide-opened eyes show only the white; her breast heaves, and a sort of death-rattle escapes from her throat.

'Go there. It is my will,' repeats the magnetizer.

'I am there!' says the unhappy girl between her teeth, falling back exhausted. Then she no longer answers; her head hangs heavy on her shoulder; her arms fall idly by her side. They approach her. They touch her. They try to waken her, but it is too late; the crime was accomplished; the woman was dead. It was to the public incredulity in the matter of magnetism that the authors of this sacrilegious experiment owed their own immunity from prosecution. The authorities held an inquest, and death was attributed to the rupture of an aneurism. The body, anyhow, bore no trace of violence; they had it buried, and there was an end of the matter.

Here is another anecdote which we heard from a travelling companion.

Two friends were staying in the same inn, and sharing the same room. One of them had a habit of talking in his sleep, and, at that time, would answer the questions which his comrade put to him. One night, he suddenly uttered stifled cries; his companion woke up and asked him what was the matter.

'But, don't you see,' said the sleeper, 'don't you see that enormous stone ... it is becoming loose from the mountain ... it is falling on me, it is going to crush me.'

'Oh, well, get out of its way!'

'Impossible! My feet are caught in brambles that cling ever closer. Ah! Help! Help! There is the great stone coming right upon me!'

'Well, there it is!' said the other laughing, throwing the pillow at his head in order to wake him.

A terrible cry, suddenly strangled in his throat, a convulsion, a sigh, then nothing more. The practical joker gets up, pulls his comrade's arm, calls him; in his turn, he becomes frightened, he cries out, people come with lights . . . the unfortunate sleep-waker is dead.

CHAPTER III

Mysteries of Hallucinations and of the Evocation of Spirits

AN HALLUCINATION is an illusion produced by an irregular movement of the astral light.

It is, as we said previously, the admixture of the phenomena of sleep with those of waking.

Our plastic medium breathes in and out the astral light or vital soul of the earth, as our body breathes in and out the terrestrial atmosphere. Now, just as in certain places the air is impure and not fit for breathing, in the same way, certain unusual circumstances may make the astral light unwholesome, and not assimilable.

The air of some places may be too bracing for some people, and suit others perfectly; it is exactly the same with the magnetic light.

The plastic medium is like a metallic statue always in a state of fusion. If the mould is defective, it becomes deformed; if the mould breaks, it runs out.

The mould of the plastic medium is balanced and polarized vital force. Our body, by means of the nervous system, attracts and retains this fugitive form of light; but local fatigue, or partial over-excitement of the apparatus, may occasion fluidic deformities.

These deformities partially falsify the mirror of the imagination, and thus occasion habitual hallucinations to the static type of visionary.

The plastic medium, made in the image and likeness of our body, of which it figures every organ in light, has a sight, touch, hearing, smell and taste which are proper to itself; it may, when it is over-excited, communicate them by vibrations to the nervous apparatus in such a manner that the hallucination is complete. The imagination seems then to triumph over Nature itself, and produces truly strange phenomena. The material body, deluged with fluid, seems to participate in the fluidic qualities, it escapes from the operation of the laws of gravity, becomes momentarily invulnerable, and even invisible, in a circle of persons suffering from collective hallucination. The convulsionaries of St. Medard, it is said, had their flesh torn off with red-hot pincers, had themselves felled like oxen, and ground like corn, and crucified, without suffering any pain; they were levitated, walked about head downwards, and ate bent pins and digested them.

We think we ought to recapitulate here the remarks which we published in the *Estafette* on the prodigies produced by the American medium Home, and on several phenomena of the same kind.

We have never personally witnessed Mr. Home's miracles, but our information comes from the best sources; we gathered it in a house where the American medium had been received with kindness when he was in misfortune, and with indulgence when he reached the point of thinking that his illness was a piece of good luck; in the house of a lady born in Poland, but thrice French by the nobility of her heart, the indescribable charm of her spirit, and the European celebrity of her name.

The publication of this information in the *Estafette* attracted to us at that time, without our particularly knowing why, the insults of a Mr. de Pène, since then become known to fame through his unfortunate duel. We thought at the time of La Fontaine's fable about the fool who threw stones at the sage. Mr. de Pène spoke of us as an unfrocked priest, and a bad Catholic. We at least showed ourself a good Christian in pitying and forgiving him, and as it is impossible to be an unfrocked priest without ever having been a priest, we let fall to the ground an insult which did not reach us.

Spooks in Paris

Mr. Home, a week ago, was once more about to quit Paris, that Paris where even the angels and the demons, if they appeared in any shape, would not pass very long for marvellous beings, and would find nothing better to do than to return at top-speed to Heaven or to hell, to escape the forgetfulness and the neglect of human kind.

Mr. Home, his air sad and disillusioned, was then bidding farewell to a noble lady whose kindly welcome had been one of the first happinesses which he had tasted in France. Mme de B—— treated him very kindly that day, as always, and asked him to stay to dinner; the man of mystery was about to accept, when some one having just said that they were waiting for a qabalist, well known in the world of occult science by the publication of a book entitled *Transcendental Magic*, Mr. Home suddenly changed countenance, and said, stammering, and with a visible embarrassment, that he could not remain, and that the approach of this Professor of Magic caused him an incomparable terror. Everything one could say to reassure him proved useless. 'I do not presume to judge the man,' said he; 'I do not assert that he is good or evil, I know nothing about it; but his atmosphere hurts me; near him I should feel myself, as it were, without force, even without life.' After which explanation, Mr. Home hastened to salute and withdraw.

This terror of miracle-mongers in the presence of the veritable initiates of science is not a new fact in the annals of occultism. You may read in Philostratus the history of the Lamia who trembled on hearing the approach of Apollonius of Tyana. Our admirable storyteller Alexandre Dumas dramatized this magical anecdote in the magnificent epitome of all legends which forms the prologue to his great epic novel, *The Wandering Jew*.[1] The scene takes place at Corinth, it is an old-time wedding with its beautiful children crowned with flowers, bearing the nuptial torches, and singing gracious epithalamia flowered with voluptuous images like the poems of Catullus. The bride is as beautiful in her chaste draperies as the ancient Polyhymnia; she is amorous and deliciously provoking in her modesty, like a Venus of Correggio, or a Grace of Canova. The bridegroom is Clinias, a disciple of the famous Apollonius of Tyana. The master had promised to come to his disciple's wedding, but he does not arrive, and the fair bride breathes easier, for she fears Apollonius. However, the day is not over. The hour has arrived when the newly married are to be conducted to the nuptial couch. Meroe trembles, pales, looks obstinately towards the door, stretches out her hand with alarm and says in a strangled voice: 'Here he is! It is he!' It was in fact Apollonius. Here is the magus; here is the master; the hour of enchantments has passed; jugglery falls before true science. One seeks the lovely bride, the white Meroe, and one sees no more than an old woman, the sorceress Canidia, the devourer of little children. Clinias is disabused; he thanks his master, he is saved.

The vulgar are always deceived about magic, and confuse adepts with enchanters. True magic, that is to say, the traditional science of the magi, is the mortal enemy of enchantment; it prevents, or makes

[1] Some authorities attribute this novel to Eugène Sue.—A.C.

to cease, sham miracles, hostile to the light, that fascinate a small number of prejudiced or credulous witnesses. The apparent disorder in the laws of Nature is a lie: it is not then a miracle. The true miracle, the true prodigy always flaming in the eyes of all, is the ever constant harmony of effect and cause; these are the splendours of eternal order!

We could not say whether Cagliostro would have performed miracles in the presence of Swedenborg; but he would certainly have dreaded the presence of Paracelsus and of Henry Khunrath, if these great men had been his contemporaries.

Far be it from us, however, to denounce Mr. Home as a low-class sorcerer, that is to say, as a charlatan. The celebrated American medium is sweet and natural as a child. He is a poor and over-sensitive being, without cunning and without defence; he is the plaything of a terrible force of whose nature he is ignorant, and the first of his dupes is certainly himself.

The study of the strange phenomena which are produced in the neighbourhood of this young man is of the greatest importance. One must seriously reconsider the too easy denials of the eighteenth century, and open out before science and reason broader horizons than those of a bourgeois criticism, which denies everything which it does not yet know how to explain to itself. Facts are inexorable, and genuine good faith should never fear to examine them.

The explanation of these facts, which all traditions obstinately affirm, and which are reproduced before our eyes with tiresome publicity, this explanation, ancient as the facts themselves, rigorous as mathematics, but drawn for the first time from the shadows in which the hierophants of all ages have hidden it, would be a great scientific event if it could obtain sufficient light and publicity. This event we are perhaps about to prepare, for one would not permit us the audacious hope of accomplishing it.

Here, in the first place, are the facts, in all their singularity. We have verified them, and we have established them with a rigorous exactitude, abstaining in the first place from all explanation and all commentary.

Mr. Home is subject to trances which put him, according to his own account, in direct communication with the soul of his mother, and, through her, with the entire world of spirits. He describes, like the sleep-wakers of Cahagnet, persons whom he has never seen, and who are recognized by those who evoke them; he will tell you even their names, and will reply, on their behalf, to questions which can be understood only by the soul evoked and yourselves.

When he is in a room, inexplicable noises make themselves heard. Violent blows resound upon the furniture, and on the walls; sometimes doors and windows open by themselves, as if they were blown open by a storm; one even hears the wind and the rain, though when one goes out of doors, the sky is cloudless, and one does not feel the lightest breath of wind.

The furniture is overturned and displaced, without anybody touching it.

Pencils write of their own accord. Their writing is that of Mr. Home, and they make the same mistakes as he does.

Those present feel themselves touched and seized by invisible hands. These contacts, which seem to select ladies, lack a serious side, and sometimes even propriety. We think that we shall be sufficiently understood.

Visible and tangible hands come out, or seem to come out, of tables; but in this case, the tables must be covered. The invisible agent needs certain apparatus, just as do the cleverest successors of Robert Houdini.

These hands show themselves above all in darkness; they are warm and phosphorescent, or cold and black. They write stupidities, or touch the piano; and when they have touched the piano, it is necessary to send for the tuner, their contact being always fatal to the exactitude of the instrument.

One of the most considerable personages in England, Sir Bulwer Lytton, has seen and touched those hands; we have read his written and signed attestation. He declares even that he has seized them, and drawn them towards himself with all his strength, in order to withdraw from their incognito the arm to which they should naturally be attached. But the invisible object has proved stronger than the English novelist, and the hands have escaped him.

A Russian nobleman who was the protector of Mr. Home, and whose character and good faith could not possibly be doubted, Count A. B——, has also seen and seized with vigour the mysterious hands. 'They are,' says he, 'perfect shapes of human hands, warm and living, *only one feels no bones.*' Pressed by an unavoidable constraint, those hands did not struggle to escape, but grew smaller, and in some way melted, so that the Count ended by no longer holding anything.

Other persons who have seen them, and touched them, say that the fingers are puffed out and stiff, and compare them to gloves of india-rubber, swollen with a warm and phosphorescent air. Sometimes, instead of hands, it is feet which produce themselves, but never naked. The spirit, which probably lacks footwear, respects (at least in this particular) the delicacy of ladies, and never shows his feet but under a drapery or a cloth.

The production of these feet very much tires and frightens Mr. Home. He then endeavours to approach some healthy person, and seizes him like a drowning man; the person so seized by the medium feels himself, on a sudden, in a singular state of exhaustion and debility.

A Polish gentleman, who was present at one of the *séances* of Mr. Home, had placed on the ground between his feet a pencil on a paper, and had asked for a sign of the presence of the spirit. For some instants nothing stirred, but suddenly, the pencil was thrown to the other end of the room. The gentleman stooped, took the paper, and saw there three qabalistic signs which nobody understood. Mr. Home (alone) appeared, on seeing them, to be very much upset, and even frightened; but he refused to explain himself as to the nature and significance of these characters. The investigators accordingly kept them, and took them to that Professor of High Magic whose approach had been so much dreaded by the medium. We have seen them, and here is a minute description of them.

They were traced forcibly, and the pencil had almost cut the paper.

They had been dashed on to the paper without order or alignment.

The first was the symbol which the Egyptian initiates usually placed in the hand of Typhon. A tau with upright double lines opened in the form of a compass; an ankh (or crux ansata) having at the top a circular ring; below the ring, a double horizontal line; beneath the double horizontal line, two oblique lines, like a V upside down.

The second character represented a Grand Hierophant's cross, with the three hierarchical cross-bars. This symbol, which dates from the remotest antiquity, is still the attribute of our sovereign pontiffs, and forms the upper extremity of their pastoral staff. But the sign traced by the pencil had this particularity, that the upper branch, the head of the cross, was double, and formed again the terrible Typhonian V, the sign of antagonism and separation, the symbol of hate and eternal combat.

The third character was that which Freemasons call the Philosophical Cross, a cross with four equal arms, with a point in each of its angles. But, instead of four points, there were only two, placed in the two right-handed corners, once more a sign of struggle, separation and denial.

The Professor, whom one will allow us to distinguish from the narrator, and to name in the third person in order not to weary our readers in having the air of speaking of ourself—the Professor, then, Master Éliphas Lévi, gave the persons assembled in Mme de B——'s drawing-room the scientific explanation of the three signatures, and this is what he said:

'These three signs belong to the series of sacred and primitive hieroglyphs, known only to initiates of the first order. The first is the signature of Typhon. It expresses the blasphemy of the evil spirit by establishing dualism in the creative principle. For the crux ansata of Osiris is a lingam upside down, and represents the paternal and active force of God (the vertical line extending from the circle) fertilizing passive nature (the horizontal line). To double the vertical line is to affirm that nature has two fathers; it is to put adultery in the place of the divine motherhood, it is to affirm, instead of the principle of intelligence, blind fatality, which has for result the eternal conflict of appearances in nothingness; it is, then, the most ancient, the most authentic, and the most terrible of all the stigmata of hell. It signifies the *atheistic god*; it is the signature of Satan.

'This first signature is hieratical, and bears reference to the occult characters of the divine world.

'The second pertains to philosophical hieroglyphs, it represents the graduated extent of idea, and the progressive extension of form.

It is a triple tau upside down; it is human thought affirming the absolute in the three worlds, and that absolute ends here by a fork, that is to say, by the sign of doubt and antagonism. So that, if the first character means: "There is no God," the rigorous signification of this one is: "Hierarchical truth does not exist."

'The third or philosophical cross has been in all initiations the symbol of Nature, and its four elementary forms. The four points represent the four indicible and incommunicable letters of the occult tetragram, that eternal formula of the Great Arcanum, G . ∴ . A . ∴ .

'The two points on the right represent force, as those on the left symbolize love, and the four letters should be read from right to left, beginning by the right-hand upper corner, and going thence to the left-hand lower corner, and so for the others, making the Cross of St. Andrew.

'The suppression of the two left-hand points expresses the negation of the Cross, the negation of mercy and of love.

'The affirmation of the absolute reign of force, and its eternal antagonism, from above to beneath, and from beneath to above.

'The glorification of tyranny and of revolt.

'The hieroglyphic sign of the unclean rite, with which, rightly or wrongly, the Templars were reproached; it is the sign of disorder and of eternal despair.'

.   .   .   .   .

Such, then, are the first revelations of the hidden science of the magi with regard to these phenomena of supernatural manifestations. Now let it be permitted to us to compare with these strange signatures other contemporary apparitions of phenomenal writings, for it is really a brief which science ought to study before taking it to the tribunal of public opinion. One must then despise no research, overlook no clue.

In the neighbourhood of Caen, at Tilly-sur-Seulles, a series of inexplicable facts occurred some years ago, under the influence of a medium, or ecstatic, named Eugène Vintras.

Certain ridiculous circumstances and a prosecution for swindling soon caused this thaumaturgist to fall into oblivion, and even into contempt; he had, moreover, been attacked with violence in pamphlets whose authors had at one time been admirers of his doctrine, for the medium Vintras took it upon himself to dogmatize. One thing, however, is remarkable in the invectives of which he is the object: his adversaries, though straining every effort in order to scourge him, recognize the truth of his miracles, and content themselves with attributing them to the devil.

What, then, are these so authentic miracles of Vintras? On this subject we are better informed than anybody, as will soon appear. Affidavits signed by honourable witnesses, persons who are artists, doctors, priests, all men above reproach, have been communicated to us; we have questioned eye-witnesses, and, better than that, we have seen with our own eyes. The facts deserve to be described in detail.

There is in Paris a writer named Mr. Madrolle, who is, to say the least of it, a bit eccentric. He is an old man of good family. He wrote at first on behalf of Catholicism in the most exalted way, received most flattering encouragements from ecclesiastical authority, and even letters from the Holy See. Then he saw Vintras; and, led away by the prestige of his miracles, became a determined sectarian, and an irreconcilable enemy of the hierarchy and of the clergy.

At the period when Éliphas Lévi was publishing his *Transcen-*

dental Magic, he received a pamphlet from Mr. Madrolle which astonished him. In it, the author vigorously sustained the most unheard-of paradoxes in the disordered style of the ecstatics. For him, life sufficed for the expiation of the greatest crimes, since it was the consequence of a sentence of death. The most wicked men, being the most unhappy of all, seemed to him to offer the sublimest of expiations to God. He broke all bounds in his attack on all repression and all damnation. 'A religion which damns,' he cried, 'is a damned religion!' He further preached the most absolute licence under the pretext of charity, and so far forgot himself as to say, that *the most imperfect and the most apparently reprehensible act of love was worth more than the best of prayers.*[1] It was the Marquis de Sade turned preacher![2] Further, he denied the existence of the devil with an enthusiasm often full of eloquence.

'Can you conceive,' said he, 'a devil tolerated and authorized by God? Can you conceive, further, a God who made the devil, and who allowed him to ravage creatures already so weak, and so prompt to deceive themselves! A god of the devil, in short, abetted, protected, and scarcely surpassed in his revenges, by a devil of a god!' The rest of the pamphlet was of the same vigour. The Professor of Magic was almost frightened, and inquired the address of Mr. Madrolle. It was not without some trouble that he obtained an interview with this singular pamphleteer, and here is, more or less, their conversation:

ÉLIPHAS LÉVI: 'Sir, I have received a pamphlet from you. I am come to thank you for your gift, and, at the same time, to testify to my astonishment and disappointment.'

MR. MADROLLE: 'Your disappointment, sir! Pray explain yourself, I do not understand you.'

'It is a lively regret to me, sir, to see you make mistakes which I have myself at one time made. But I had then, at least, the excuse of inexperience and youth. Your pamphlet lacks conviction, because it lacks discrimination. Your intention was doubtless to protest against errors in belief, and abuses in morality: and behold, it is the belief and the morality themselves that you attack! The exaltation which overflows in your pamphlet may indeed do you the greatest harm, and

---

[1] Quoted with approval in *Solution of the First Problem,* IX, p. 41.—A.C. It is difficult to determine whether the words 'act of love' should be interpreted in their gross, or in their mystical, sense. Perhaps Madrolle was himself intentionally ambiguous.—A.C.

[2] But the Marquis de Sade was, above all, a preacher. Three-fourths of *Justine* are verbose arguments in favour of vice. Again Lévi trips in referring to an author whom he has not read.—A.C.

some of your best friends must have experienced anxiety with regard to the state of your health. . . .'

'Oh, no doubt; they have said, and say still, that I am mad. But it is nothing new that believers must undergo the folly of the Cross. I am exalted, sir, because you yourself would be so in my place, because it is impossible to remain calm in the presence of prodigies. . . .'

'Oh, oh, you speak of prodigies, that interests me. Come, between ourselves, and in all good faith, of what prodigies are you speaking?'

'Eh, what prodigies should they be but those of the great prophet Elias, returned to earth under the name of Pierre Michel?'

'I understand; you mean Eugène Vintras. I have heard his prophecies spoken of. But does he really perform miracles?'

(*Here Mr. Madrolle jumps in his chair, raises his eyes and his hands to Heaven, and finally smiles with a condescension which seems to sound the depths of pity.*)

'Does he do miracles, sir?

'But the greatest!

'The most astonishing!

'The most incontestable!

'The truest miracles that have ever been done on earth since the time of Jesus Christ! . . . What! Thousands of hosts appear on altars where there were none; wine appears in empty chalices, and it is not an illusion, it is wine, a delicious wine . . . celestial music is heard, perfumes of the world beyond fill the room, and then blood . . . real human blood (doctors have examined it!), real blood, I tell you, sweats and sometimes flows from the hosts, imprinting mysterious characters on the altars! I am talking to you of what I have seen, of what I have heard, of what I have touched, of what I have tasted! And you want me to remain cold at the bidding of an ecclesiastical authority which finds it more convenient to deny everything than to examine the least thing! . . .'

'By permission, sir; it is in religious matters, above all, that authority can never be wrong. . . . In religion, good is hierarchy, and evil is anarchy; to what would the influence of the priesthood be reduced, in effect, if you set up the principle that one must rather believe the testimony of one's senses than the decision of the Church? Is not the Church more visible than all your miracles? Those who see miracles and who do not see the Church are much more to be pitied than the blind, for there remains to them not even the resource of allowing themselves to be led. . . .'

'Sir, I know all that as well as you do. But God cannot be divided against Himself. He cannot allow good faith to be deceived, and the Church itself could hardly decide that I am blind when I have eyes. . . . Here, see what John Huss says in his letter, the forty-third letter, towards the end:

' "A doctor of theology said to me: 'In everything I should submit myself to the Council; everything would then be good and lawful for me.' He added: 'If the Council said that you had only one eye, although you have two, it would be still necessary to admit that the Council was not wrong.' 'Were the whole world,' I replied, 'to affirm such a thing, so long as I had the use of my reason, I should not be able to agree without wounding my conscience.' " I will say to you, like John Huss, "Before there were a Church and its councils there were truth and reason." '

'Pardon me if I interrupt, my dear sir; you were a Catholic at one time, you are no longer so; consciences are free. I shall merely submit to you that the institution of the hierarchical infallibility in matters of dogma is reasonable in quite another sense, and far more incontestably true than all the miracles of the world. Besides, what sacrifices ought one not to make in order to preserve peace! Believe me, John Huss would have been a greater man if he had sacrificed one of his eyes to universal concord, rather than deluge Europe with blood! O sir! let the Church decide when she will that I have but one eye; I only ask her one favour, it is to tell me in which eye I am blind, in order that I may close it and look with the other with an irreproachable orthodoxy!'

'I admit that I am not orthodox in your fashion.'

'I perceive that clearly. But let us come to the miracles! You have then seen, touched, felt, tasted them; but, come, putting exaltation on one side, please give me a thoroughly detailed and circumstantial account of the affair, and, above all, evident proof of miracle. Am I indiscreet in asking you that?'

'Not the least in the world; but which shall I choose? There are so many!'

'Let me think,' added Mr. Madrolle, after a moment's reflection and with a slight trembling in the voice, 'the prophet is in London, and we are here. Eh! well, if you only make a mental request to the prophet to send you immediately the communion, and if in a place designated by you, in your own house, in a cloth, or in a book, you found a host on your return, what would you say?'

'I should declare the fact inexplicable by ordinary critical rules.'

'Oh, well, sir,' cried Madrolle, triumphantly, there is a thing that often happens to me; whenever I wish, that is to say, whenever I am prepared and hope humbly to be worthy of it! Yes, sir, I find the host when I ask for it; I find it real and palpable, but often ornamented with little hearts, little miraculous hearts, which one might think had been painted by Raphael.'

Éliphas Lévi, who felt ill at ease in discussing facts with which there was mingled a sort of profanation of the most holy things, then took his leave of the one-time Catholic writer, and went out meditating on the strange influence of this Vintras, who had so overthrown that old belief, and turned the old savant's head.

Some days afterwards, the qabalist Éliphas was awakened very early in the morning by an unknown visitor. It was a man with white hair, entirely clothed in black; his physiognomy that of an extremely devout priest; his whole air, in short, was entirely worthy of respect.

This ecclesiastic was furnished with a letter of recommendation conceived in these terms:

Dear Master,
   This is to introduce to you an old savant, who wants to gabble Hebrew sorcery with you. Receive him like myself—I mean as I myself received him—by getting rid of him in the best way you can.
      Entirely yours, in the sacrosanct Qabalah,
          AD. DESBARROLLES.

'Reverend sir,' said Éliphas, smiling, after having read the letter, 'I am entirely at your service, and can refuse nothing to the friend who writes to me. You have then seen my excellent disciple Desbarrolles?'

'Yes, sir, and I have found in him a very amiable and very learned man. I think both you and him worthy of the truth which has been lately revealed by astonishing miracles, and the positive revelations of the Archangel St. Michael.'

'Sir, you do us honour. Has then the good Desbarrolles astonished you by his science?'

'Oh, certainly he possesses in a very remarkable degree the secrets of cheiromancy; by merely inspecting my hand, he told me nearly the whole history of my life.'

'He is quite capable of that. But did he enter into the smallest details?'

'Sufficiently, sir, to convince me of his extraordinary power.'

'Did he tell you that you were once the vicar of Mont-Louis, in the diocese of Tours? That you are the most zealous disciple of the ecstatic Eugène Vintras? And that your name is Charvoz?'

It was a veritable thunderbolt; at each of these three phrases the old priest jumped in his chair. When he heard his name, he turned pale, and rose as if a spring had been released.

'You are then really a magician?' he cried; 'Charvoz is certainly my name, but it is not that which I bear; I call myself La Paraz.'

'I know it; La Paraz is the name of your mother. You have left a sufficiently enviable position, that of a country vicar, and your charming vicarage, in order to share the troubled existence of a sectary.'

'Say of a great prophet!'

'Sir, I believe perfectly in your good faith. But you will permit me to examine a little the mission and the character of your prophet.'

'Yes, sir; examination, full light, the microscope of science, that is all we ask. Come to London, sir, and you will see! The miracles are permanently established there.'

'Would you be so kind, sir, as to give me, first of all, some exact and conscientious details with regard to the miracles?'

'Oh, as many as you like!'

And immediately the old priest began to recount things which the whole world would have found impossible, but which did not even turn an eye-lash of the Professor of Transcendental Magic.

Here is one of his stories:

One day Vintras, in an access of enthusiasm, was preaching before his heterodox altar; twenty-five persons were present. An empty chalice was upon the altar, a chalice well known to the Abbé Charvoz; he brought it himself from his church of Mont-Louis, and he was perfectly certain that the sacred vase had neither secret ducts nor double bottom.

' "In order to prove to you," said Vintras, "that it is God Himself who inspires me, He acquaints me that this chalice will fill itself with drops of His blood, under the appearance of wine, and you will all be able to taste the fruit of the vines of the future, the wine which we shall drink with the Saviour in the Kingdom of His Father. . . ."

'Overcome with astonishment and fear,' continued the Abbé Charvoz, 'I went up to the altar, I took the chalice, I looked at the bottom of it: it was entirely empty. I overturned it in the sight of every-one, then I returned to kneel at the foot of the altar, holding the chalice between my two hands. . . . Suddenly there was a slight noise; the noise of a drop of water, falling into the chalice from the ceiling, was distinctly heard, and a drop of wine appeared at the bottom of the vase.

'Every eye was fixed on me. Then they looked at the ceiling, for our simple chapel was held in a poor room; in the ceiling was neither hole nor fissure; nothing was seen to fall, and yet the noise of the fall of the drops multiplied, it became more rapid, and more frequent . . . and the wine climbed from the bottom of the chalice towards the brim.

'When the chalice was full, I bore it slowly around so that all might see it; then the prophet dipped his lips into it, and all, one after the other, tasted the miraculous wine. It is in vain to search memory for any delicious taste which would give an idea of it. . . . And what shall I tell you,' added the Abbé Charvoz, 'of those miracles of blood which astonish us every day? Thousands of wounded and bleeding hosts are found upon our altars. The sacred stigmata appear to all who wish to see them. The hosts, at first white, slowly become marked with characters and hearts in blood. . . . Must one believe that God abandons the holiest objects to the false miracles of the devil? Should not one rather adore, and believe that the hour of the supreme and final revelation has arrived?'

Abbé Charvoz, as he thus spoke, had in his voice that sort of nervous trembling that Éliphas Lévi had already noticed in the case of Mr. Madrolle. The magician shook his head pensively; then, suddenly:

'Sir,' said he to the Abbé; 'you have upon you one or two of these miraculous hosts. Be good enough to show them to me.'

'Sir——'

'You have some, I know it; why should you deny it?'

'I do not deny it,' said Abbé Charvoz; 'but you will permit me not to expose to the investigations of incredulity objects of the most sincere and devout belief.'

'Reverend sir,' said Éliphas gravely; 'incredulity is the mistrust of an ignorance almost sure to deceive itself. Science is not incredulous. I believe, to begin with, in your own conviction, since you have accepted a life of privation and even of reproach, in order to stick to this unhappy belief. Show me then your miraculous hosts, and believe entirely in my respect for the objects of a sincere worship.'

'Oh, well!' said the Abbé Charvoz, after another slight hesitation; 'I will show them to you.'

Then he unbuttoned the top of his black waistcoat and drew forth a little reliquary of silver, before which he fell on his knees, with tears in his eyes, and prayers on his lips; Éliphas fell on his knees beside him, and the Abbé opened the reliquary.

There were in the reliquary three hosts, one whole, the two others almost like paste, and as it were kneaded with blood.

The whole host bore in its centre a heart in relief on both sides; a clot of blood moulded in the form of a heart, which seemed to have been formed in the host itself in an inexplicable manner. The blood could not have been applied from without, for the imbibed colouring matter had left the particles adhering to the exterior surface quite white. The appearance of the phenomenon was the same on both sides. The Master of Magic was seized with an involuntary trembling.

This emotion did not escape the old vicar, who having once again done adoration and closed his reliquary, drew from his pocket an album, and gave it without a word to Éliphas. . . . There were copies of all the bleeding characters which had been observed upon hosts since the beginning of the ecstasies and miracles of Vintras.

There were hearts of every kind, and many different sorts of emblems. But three especially excited the curiosity of Éliphas, to the highest point.

'Reverend sir,' said he to Charvoz, 'do you know these three signs?'

'No,' replied the Abbé ingenuously; 'but the prophet assures us that they are of the highest importance, and that their hidden significa-tion shall soon be made known, that is to say, at the end of the Age.'

'Oh, well, sir,' solemnly replied the Professor of Magic; 'even before the end of the Age, I will explain them to you: these three qabalistic signs are the signature of the devil!'

'It is impossible!' cried the old priest.

'It is the case,' replied Éliphas, with determination.

Now, the signs were these:

1°. The star of the microcosm, or the magic pentagram. It is the five-pointed star of occult masonry, the star with which Agrippa drew the human figure, the head in the upper point, the four limbs in the four others. The flaming star, which, when turned upside down, is the hieroglyphic sign of the goat of Black Magic, whose head may then be drawn in the star, the two horns at the top, the ears to the right and left, the beard at the bottom. It is the sign of antagonism and fatality. It is the goat of lust attacking the Heavens with its horns. It is a sign execrated by initiates of a superior rank, even at the Sabbath.[1]

[1] But if this were on a circular host, how could it be upside down?—A.C.

2°. The two hermetic serpents. But the heads and tails, instead of coming together in two similar semicircles, were turned outwards, and there was no intermediate line representing the caduceus. Above the head of the serpents, one saw the fatal V, the Typhonian fork, the character of hell. To the right and left, the sacred numbers III and VII were relegated to the horizontal line which represents passive and secondary things. The meaning of the character was then this:

Antagonism is eternal.

God is the strife of fatal forces, which always create through destruction.

The things of religion are passive and transitory.

Boldness makes use of them, war profits by them, and it is by them that discord is perpetuated.

3°. Finally, the qabalistic monogram of Jehovah, the JOD and the HÉ, but upside down. This, according to the doctors of occult science, is the most frightful of all blasphemies, and signifies, however one may read it, 'Fatality alone exists: God and the Spirit are not. Matter is all, and spirit is only a fiction of this matter demented. Form is more than idea, woman more than man, pleasure more than thought, vice more than virtue, the mob more than its chiefs, the children more than their fathers, folly more than reason!'

There is what was written in characters of blood upon the pretended miraculous hosts of Vintras!

We affirm upon our honour that the facts cited above are such as we have stated, and that we ourselves saw and explained the characters according to magical science and the true keys of the Qabalah.

The disciple of Vintras also communicated to us the description and design of the pontifical vestments given, said he, by Jesus Christ Himself to the pretended prophet, during one of his ecstatic trances. Vintras had these vestments made, and clothes himself with them in order to perform his miracles. They are red in colour. He wears upon his forehead a cross in the form of a lingam; and his pastoral staff is surmounted by a hand, all of whose fingers are closed, except the thumb and the little finger.

Now, all that is diabolical in the highest degree. And is it not a really wonderful thing, this intuition of the signs of a lost science? For it is transcendental magic which, basing the universe upon the two columns of Hermes and of Solomon, has divided the metaphysical world into two intellectual zones, one white and luminous, enclosing positive ideas, the other black and obscure containing negative ideas, and which has given to the synthesis of the first, the name of God, and to that of the other, the name of the devil or of Satan.

The sign of the lingam borne upon the forehead is in India the distinguishing mark of the worshippers of Shiva the destroyer; for that sign being that of the great magical arcanum, which refers to the mystery of universal generation, to bear it on the forehead is to make profession of dogmatic shamelessness. 'Now,' say the Orientals, 'the day when there is no longer modesty in the world, the world, given over to debauch which is sterile, will end at once for lack of mothers. Modesty is the acceptance of maternity.'

The hand with the three large fingers closed expresses the negation of the ternary, and the affirmation of the natural forces alone.

The ancient hierophants, as our learned and witty friend Desbarrolles is about to explain in an admirable book which is at present in the press, had given a complete *résumé* of magical science in the human hand. The forefinger, for them, represented Jupiter; the middle finger, Saturn; the ring-finger, Apollo or the Sun. Among the Egyptians, the middle finger was Ops, the forefinger Osiris, and the little finger Horus; the thumb represented the generative force, and the little finger, cunning. A hand, showing only the thumb and the little finger, is equivalent, in the sacred hieroglyphic language, to the exclusive affirmation of passion and diplomacy. It is the perverted and material translation of that great word of St. Augustine: 'Love, and do what you will!' Compare now this sign with the doctrine of Mr. Madrolle: *The most imperfect and the most apparently guilty act of love is worth more than the best of prayers.* And you will ask yourself what is that force which, independently of the will, and of the greater or less knowledge of man (for Vintras is a man of no education), formulates its dogmas with signs buried in the rubbish of the ancient world, re-discovers the mysteries of Thebes and of Eleusis, and writes for us the most learned reveries of India with the occult alphabets of Hermes?

What is that force? I will tell you. But I have still plenty of other miracles to tell; and this article is like a judicial investigation. We must, before anything else, complete it.

However, we may be permitted, before proceeding to other accounts, to transcribe here a page from a German *illuminé*, the work of Ludwig Tieck:

If, for example, as an ancient tradition informs us, some of the angels whom God had created fell all too soon, and if these, as they also say, were precisely the most brilliant of the angels, one may very well understand by this 'fall' that they sought a new road, a new form of activity, other occupations, and another life than those orthodox or more passive spirits who remained in the realm assigned to them, and made no use of liberty, the appanage of all of them. Their 'fall' was that weight of form which we now-a-days call reality, and which is a protest on the part of individual existence against its reabsorption into the abysses of universal spirit. It is thus that death preserves and reproduces life, it is thus that life is betrothed to death. ... Do you understand now what Lucifer is? Is it not the very genius of ancient Prometheus, that force which sets in motion the world, life, even movement, and which regulates the course of successive forms? This force, by its resistance, equilibrated the creative principle. It is thus that the Elohim gave birth to the earth. When, subsequently, men were placed upon the earth by the Lord, as intermediate spirits, in their enthusiasm, which led them to search Nature in its depths, they gave themselves over to the influence of that proud and powerful genius, and when they were softly ravished away over the precipice of death to find life, there it was that they began to exist in a real and natural manner as is fit for all creatures.

This page needs no commentary, and explains sufficiently the tendencies of what one calls spiritualism, or *spiritism.*

It is already a long time since this doctrine, or, rather, this anti-doctrine, began to work upon the world, to plunge it into universal anarchy. But the law of equilibrium will save us, and already the great movement of reaction has begun.

We continue the recital of the phenomena.

One day a workman paid a visit to Éliphas Lévi. He was a tall man of some fifty years, of frank appearance, and speaking in a very reasonable manner. Questioned as to the motive of his visit, he replied: 'You ought to know it well enough; I am come to beg and pray you to return to me what I have lost.'

We must say, to be frank, that Éliphas knew nothing of this visitor, nor of what he might have lost. He accordingly replied: 'You think me much more of a sorcerer than I am; I do not know who you are, nor what you seek; consequently, if you think that I can be useful to you in any way, you must explain yourself and make your request more precise.'

'Oh, well, since you are determined not to understand me, you will at least recognize this,' said the stranger, taking from his pocket a little, much-used black book.

It was the *grimoire* of Pope Honorius.

One word upon the little book so much decried.

The *grimoire* of Honorius is composed of an apocryphal constitution of Honorius II, for the evocation and control of spirits; then of some superstitious receipts . . . it was the manual of the bad priests who practised Black Magic during the darkest periods of the middle ages. You will find there bloody rites, mingled with profanations of the Mass and of the consecrated elements, formulae of bewitchment and malevolent spells, and practices which stupidity alone could credit or knavery counsel. In fact, it is a book complete of its kind; it is consequently become very rare, and the bibliophile pushes it to very high prices in the public sales.

'My dear sir,' said the workman, sighing, 'since I was ten years old, I have not missed once performing the orison. This book never leaves me, and I comply rigorously with all the prescribed ceremonies. Why, then, have those who used to visit me abandoned me? Éli, Éli, lama——'

'Stop,' said Éliphas, 'do not parody the most formidable words that agony ever uttered in this world! Who are the beings who visited you by virtue of this horrible book? Do you know them? Have you promised them anything? Have you signed a pact?'

'No,' interrupted the owner of the *grimoire*; I do not know them, and I have entered into no agreement with them. I only know that among them the chiefs are good, the intermediate rank partly good and partly evil; the inferiors bad, but blindly, and without its being possible for them to do better. He whom I evoked, and who has often appeared to me, belongs to the most elevated hierarchy; for he was good-looking, well dressed, and always gave me favourable answers. But I have lost a page of my *grimoire*, the first, the most important, that which bore the autograph of the spirit; and, since then, he no longer appears when I call him.

'I am a lost man. I am naked as Job, I have no longer either force or courage. O Master, I conjure you, you who need only say one word, make one sign, and the spirits will obey, take pity upon me, and restore to me what I have lost!'

'Give me your *grimoire*!' said Éliphas. 'What name used you to give to the spirit who appeared to you?'

'I called him Adonai.'

'And in what language was his signature?'

'I do not know, but I suppose it was in Hebrew.'

'There,' said the Professor of Transcendental Magic, after having traced two words in the Hebrew language in the beginning and at the end of the book. 'Here are two words which the spirits of darkness will never counterfeit. Go in peace, sleep well, and no longer evoke spirits.'

The workman withdrew.

A week later he returned to seek the Man of Science.

'You have restored to me hope and life,' said he; 'my strength is partially returned, I am able with the signatures that you gave me to relieve sufferers, and cast out devils, but *him*, I cannot see him again, and, until I have seen him, I shall be sad to the day of my death. Formerly, he was always near me, he sometimes touched me, and he used to wake me up in the night to tell me all that I needed to know. Master, I beg of you, let me see him again!'

'See whom?'

'Adonai.'

'Do you know who Adonai is?'

'No, but I want to see him again.'

'Adonai is invisible.'

'I have seen him.'

'He has no form.'

'I have touched him.'

'He is infinite.'

'He is very nearly of my own height.'

'The prophets say of him that the hem of his vestment, from the East to the West, sweeps the stars of the morning.'

'He had a very clean surcoat, and very white linen.'

'The Holy Scripture says that one cannot see him and live.'

'He had a kind and jovial face.'

'But how did you proceed in order to obtain these apparitions?'

'Why, I did everything that it tells you to do in the *grimoire*.'

'What! Even the bloody sacrifice?'

'Doubtless.'

'Unhappy man! But who, then, was the victim?'

At this question, the workman had a slight trembling; he paled, and his glance became troubled.

'Master, you know better than I what it is,' said he humbly in a low voice. 'Oh, it cost me a great deal to do it; above all, the first time, with a single blow of the magic knife to cut the throat of that innocent creature! One night I had just accomplished the funereal rites, I was seated in the circle on the interior threshold of my door, and the victim had just been consumed in a great fire of alder and cypress wood. . . .

All of a sudden, quite close to me ... I dreamt or rather I felt it pass ... I heard in my ear a heartrending wail ... one would have said that it wept; and since that moment, I think that I am hearing it always.'

Éliphas had risen; he looked fixedly upon his interlocutor. Had he before him a dangerous madman, capable of renewing the atrocities of the seigneur of Retz? And yet the face of the man was gentle and honest. No, it was not possible.

'But then this victim ... tell me clearly what it was. You suppose that I know already. Perhaps I do know, but I have reasons for wishing you to tell me.'

'It was, according to the magic ritual, a young goat of a year old, virgin, and without defect.'

'A real young he-goat?'

'Doubtless. Understand that it was neither a child's toy, nor a stuffed animal.'

Éliphas breathed again.

'Good,' thought he; 'This man is not a sorcerer worthy of the stake. He does not know that the abominable authors of the *grimoire*, when they spoke of the "virgin he-goat", meant a little child.'

'Well,' said he to his consultant; 'give me some details about your visions. What you tell me interests me in the highest degree.'

The sorcerer—for one must call him so—the sorcerer then told him of a series of strange facts, of which two families had been witness, and these facts were precisely identical with the phenomena of Mr. Home: hands coming out of walls, movements of furniture, phosphorescent apparitions. One day the rash apprentice-magician had dared to call up Astaroth, and had seen the apparition of a gigantic monster having the body of a hog, and the head borrowed from the skeleton of a colossal ox. But he told all that with an accent of truth, a certainty of having seen, which excluded every kind of doubt as to the good faith and the entire conviction of the narrator. Éliphas, who is an epicure in magic, was delighted with this find. In the nineteenth century, a real sorcerer of the middle ages, a remarkably innocent and convinced sorcerer, a sorcerer who had seen Satan under the name of Adonai, Satan dressed like a respectable citizen, and Astaroth in his true diabolical form! What a supreme find for a museum! What a treasure for an archaeologist!

'My friend,' said he to his new disciple, 'I am going to help you to find what you say you have lost. Take my book, observe the prescriptions of the ritual, and come again to see me in a week.'

A week later he returned, but this time the workman declared that he had invented a life-saving machine of the greatest importance for the navy. The machine is perfectly put together; it only lacks one thing—it will not work: there is a hidden defect in the machinery. What was that defect? The evil spirit alone could tell him. It is then absolutely necessary to evoke him! . . .

'Take care you do not!' said Éliphas. 'You had much better say for nine days this qabalistic evocation.' He gave him a leaf covered with manuscript. 'Begin this evening, and return tomorrow to tell me what you have seen, for tonight you will have a manifestation.'

The next day, our good man did not miss the appointment.

'I woke up suddenly,' said he, 'upon one o'clock in the morning. In front of my bed I saw a bright light, and in this light a *shadowy arm* which passed and repassed before me, as if to magnetize me. Then I went to sleep again, and some instants afterwards, waking anew, I saw again the same light, but it had changed its place. It had passed from left to right, and upon a luminous background I distinguished the silhouette of a man who was looking at me with arms crossed.'

'What was this man like?'

'Just about your height and breadth.'

'It is well. Go, and continue to do what I told you.'

The nine days rolled by; at the end of that time, a new visit; but this time he was absolutely radiant and excited. As soon as he caught sight of Éliphas:

'Thanks, Master!' he cried. 'The machine works! People whom I did not know have come to place at my disposal the funds which were necessary to carry out my enterprise; I have found again peace in sleep; and all that thanks to your power!'

'Say, rather, thanks to your faith and your docility. And now, fare-well: I must work. . . . Well, why do you assume this suppliant air, and what more do you want of me?'

'Oh, if you only would——'

'Well, what now? Have you not obtained all that you asked for, and even more than you asked for, for you did not mention money to me?'

'Yes, doubtless,' said the other sighing; but I do want to see him again!'

'Incorrigible!' said Éliphas.

Some days afterwards the Professor of Transcendental Magic was awakened, about two o'clock in the morning, by an acute pain in the head. For some moments he feared a cerebral congestion. He therefore rose, relit his lamp, opened his window, walked to and fro in his study, and then, calmed by the fresh air of the morning, he lay down again and slept deeply. He had a nightmare: he saw, terribly real, the giant with the fleshless ox's head of which the workman had spoken to him. The monster pursued him, and struggled with him. When he woke up, it was already day, and somebody was knocking at his door. Éliphas rose, threw on a dressing-gown, and opened; it was the workman.

'Master,' said he, entering hastily, and with an alarmed air; 'how are you?'

'Very well,' replied Éliphas.

'But last night, at two o'clock in the morning, did you not run a great danger?'

Éliphas did not grasp the allusion; he already no longer remembered the indisposition of the night.

'A danger?' said he. 'No; none that I know of.'

'Have you not been assaulted by a monster phantom, who sought to strangle you? Did it not hurt you?'

Éliphas remembered.

'Yes,' said he, 'certainly, I had the beginning of a sort of apoplectic attack, and a horrible dream. But how do you know that?'

'At the same time, an invisible hand struck me roughly on the shoulder, and awoke me suddenly. I dreamt then that I saw you fighting with Astaroth. I jumped up, and a voice said in my ear: "Arise and go to the help of thy Master; he is in danger." I got up in a great hurry. But where must I run? What danger threatened you? Was it at your own house, or elsewhere? The voice said nothing about that. I decided to wait for sunrise; and immediately day dawned, I ran, and here I am.'

'Thanks, friend,' said the magus, holding out his hand; 'Astaroth is a stupid joker; all that happened last night was a little blood to the head. Now, I am perfectly well. Be assured, then, and return to your work.'

Strange as may be the facts which we have just related, there remains for us to unveil a tragic drama much more extraordinary still.

It refers to the deed of blood which at the beginning of this year plunged Paris and all Christendom into mourning and stupefaction; a deed in which no one suspected that Black Magic had any part.

Here is what happened:

During the winter, at the beginning of last year, a bookseller informed the author of *Transcendental Magic* that an ecclesiastic was looking for his address, testifying the greatest desire to see him. Éliphas Lévi did not feel himself immediately prepossessed with confidence towards the stranger, to the point of exposing himself without precaution to his visits; he indicated the house of a friend, where he was to be in the company of his faithful disciple, Desbarrolles. At the hour and date appointed they went, in fact, to the house of Mme A——, and found that the ecclesiastic had been waiting for them for some moments.

He was a young and slim man; he had an arched and pointed nose, with dull blue eyes. His bony and projecting forehead was rather broad than high, his head was dolichocephalic, his hair flat and short, parted on one side, of a greyish blond with just a tinge of chestnut of a rather curious and disagreeable shade. His mouth was sensual and quarrelsome; his manners were affable, his voice soft, and his speech sometimes a little embarrassed. Questioned by Éliphas Lévi concerning the object of his visit, he replied that he was on the look-out for the *grimoire* of Honorius, and that he had come to learn from the Professor of Occult Science how to obtain that little black book, nowadays almost impossible to find.

'I would gladly give a hundred francs for a copy of that *grimoire*,' said he.

'The work in itself is valueless,' said Éliphas. 'It is a pretended constitution of Honorius II, which you will find perhaps quoted by some erudite collector of apocryphal constitutions; you can find it in the library.'

'I will do so, for I pass almost all my time in Paris in the public libraries.'

'You are not occupied in the ministry in Paris?'

'No, not now; I was for some little while employed in the parish of St. Germain-Auxerrois.'

'And you now spend your time, I understand, in curious researches in occult science.'

'Not precisely, but I am seeking the realization of a thought. . . . I have something to do.'

'I do not suppose that this something can be an operation of Black Magic. You know as well as I do, reverend sir, that the Church has always condemned, and still condemns, severely, everything which relates to these forbidden practices.'

A pale smile, imprinted with a sort of sarcastic irony, was all the answer that the Abbé gave, and the conversation fell to the ground.

However, the cheiromancer Desbarrolles was attentively looking at the hand of the priest; he perceived it, a quite natural explanation followed, the Abbé offered graciously and of his own accord his hand to the experimenter. Desbarrolles knit his brows, and appeared embarrassed. The hand was damp and cold, the fingers smooth and spatulated; the mount of Venus, or the part of the palm of the hand which corresponds to the thumb, was of a noteworthy development, the line of life was short and broken, there were crosses in the centre of the hand, and stars upon the mount of the moon.

'Reverend sir,' said Desbarrolles, 'if you had not a very solid religious education you would easily become a dangerous sectary, for you are led on the one hand towards the most exalted mysticism, and on the other to the most concentrated obstinacy combined with the greatest secretiveness that can possibly be. You want much, but you imagine more, and as you confide your imaginations to nobody, they might attain proportions which would make them veritable enemies for yourself. Your habits are contemplative and rather easy-going, but it is a somnolence whose awakenings are perhaps to be dreaded. You are carried away by a passion which your state of life—— But pardon, reverend sir, I fear that I am over-stepping the boundaries of discretion.'

'Say everything, sir; I am willing to hear all, I wish to know everything.'

'Oh, well! If, as I do not doubt to be the case, you turn to the profit of charity all the restless activities with which the passions of your heart furnish you, you must often be blessed for your good works.'

The Abbé once more smiled that dubious and fatal smile which gave so singular an expression to his pallid countenance. He rose and took his leave without having given his name, and without any one having thought to ask him for it.

Éliphas and Desbarrolles reconducted him as far as the staircase, in token of respect for his dignity as a priest.

Near the staircase he turned and said slowly:

'Before long, you will hear something. . . . You will hear me spoken of,' he added, emphasizing each word. Then he saluted with head and hand, turned without adding a single word, and descended the staircase.

The two friends returned to Mme A——'s room.

'There is a singular personage,' said Éliphas; 'I think I have seen Pierrot of the Funambules playing the part of a traitor. What he said to us on his departure seemed to me very much like a threat.'

'You frightened him,' said Mme A——. 'Before your arrival, he was beginning to open his whole mind, but you spoke to him of conscience and of the laws of the Church, and he no longer dared to tell you what he wished.'

'Bah! What did he wish then?'

'To see the devil.'

'Perhaps he thought I had him in my pocket?'

'No, but he knows that you give lessons in the Qabalah, and in magic, and so he hoped that you would help him in his enterprise. He told my daughter and myself that in his vicarage in the country, he had already made one night an evocation of the devil by the help of a popular *grimoire*. "Then," said he, "a whirlwind seemed to shake the vicarage; the rafts groaned, the wainscoting cracked, the doors shook, the windows opened with a crash, and whistlings were heard in every corner of the house." He then expected that formidable vision to follow but he saw nothing; no monster presented itself; in a word, the devil would not appear. That is why he is looking for the *grimoire* of Honorius, for he hopes to find in it stronger conjurations, and more efficacious rites.'

'Really! But the man is then a monster, or a madman!'

'I think he is just simply in love,' said Desbarrolles. 'He is gnawed by some absurd passion, and hopes for absolutely nothing unless he can get the devil to interfere.'

'But how then—what does he mean when he says that we shall hear him spoken of?'

'Who knows? Perhaps he thinks to carry off the Queen of England, or the Sultana Valide.'

The conversation dropped, and a whole year passed without Mme A——, or Desbarrolles, or Éliphas hearing the unknown young priest spoken of.

In the course of the night between the 1st and 2nd of January 1857, Éliphas Lévi was awakened suddenly by the emotions of a bizarre and dismal dream. It seemed to him that he was in a dilapidated room of gothic architecture, rather like the abandoned chapel of an old castle. A door hidden by a black drapery opened on to this room; behind the drapery one guessed the hidden light of tapers, and it seemed to Éliphas that, driven by a curiosity full of terror, he was approaching the black drapery. . . . Then the drapery was parted, and a hand was stretched forth and seized the arm of Éliphas. He saw no one, but he heard a low voice which said in his ear:

'Come and see your father, who is about to die.'

The magus awoke, his heart palpitating, and his forehead bathed in sweat.

'What can this dream mean?' thought he. 'It is long since my father died; why am I told that he is going to die, and why has this warning upset me?'

The following night, the same dream recurred with the same circumstances; once more Éliphas awoke, hearing a voice in his ear repeat:

'Come and see your father, who is about to die.'

This repeated nightmare made a painful impression upon Éliphas: he had accepted, for the 3rd of January, an invitation to dinner in pleasant company, but he wrote and excused himself, feeling himself little inclined for the gaiety of a banquet of artists. He remained, then, in his study; the weather was cloudy; at midday he received a visit from one of his magical pupils, Viscount M———. When he left, the rain was falling in such abundance that Éliphas offered his umbrella to the Viscount, who refused it. There followed a contest of politeness, of which the result was that Éliphas went out to see the Viscount home. While they were in the street, the rain stopped, the Viscount found a carriage, and Éliphas, instead of returning to his house, mechanically crossed the Luxembourg, went out by the gate which opens on the Rue d'Enfer, and found himself opposite the Panthéon.

A double row of booths, improvised for the Festival of St. Geneviève, indicated to pilgrims the road to St. Étienne-du-Mont. Éliphas, whose heart was sad, and consequently disposed to prayer, followed that way and entered the church. It might have been at that time about four o'clock in the afternoon.

The church was full of the faithful, and the office was performed with great concentration, and extraordinary solemnity. The banners of the parishes of the city, and of the suburbs, bore witness to the public veneration for the virgin who saved Paris from famine and invasion. At the bottom of the church, the tomb of St. Geneviève shone gloriously with light. They were chanting the litanies, and the procession was coming out of the choir.

After the Cross, accompanied by its acolytes, and followed by the choirboys, came the banner of St. Geneviève; then, walking in double file, came the lady devotees of St. Geneviève, clothed in black, with a white veil on the head, a blue ribbon around the neck with the medal of the legend, a taper in the hand, surmounted by the little gothic lantern that tradition gives to the images of the saint. For, in the old books, St. Geneviève is always represented with a medal on her neck, that which St. Germain d'Auxerre gave her, and holding a taper, which the devil tries to extinguish, but which is protected from the breath of the unclean spirit by a miraculous little tabernacle.

After the lady devotees came the clergy; then finally appeared the venerable Archbishop of Paris, mitred with a white mitre, wearing a cope which was supported on each side by his two vicars; the prelate, leaning on his cross, walked slowly, and blessed to right and left the crowd which knelt about his path. Éliphas saw the Archbishop for the first time, and noticed the features of his countenance. They expressed kindliness and gentleness; but one might observe the expression of a great fatigue, and even of a nervous suffering painfully dissimulated.

The procession descended to the foot of the church, traversing the nave, went up again by the aisle at the left of the door, and came to the station of the tomb of St. Geneviève; then it returned by the right-hand aisle, chanting the litanies as it went. A group of the faithful followed the procession, and walked immediately behind the Archbishop.

Éliphas mingled in this group, in order more easily to get through the crowd which was about to reform, so that he might regain the door of the church. He was lost in reverie, softened by this pious solemnity.

The head of the procession had already returned to the choir, the Archbishop was arriving at the railing of the nave: there the passage was too narrow for three people to walk in file; the Archbishop was in front, and the two grand-vicars behind him, always holding the edges of his cope, which was thus thrown off, and drawn backwards, in such a manner that the prelate presented his breast uncovered, and protected only by the crossed embroideries of his stole.

Then those who were behind the Archbishop saw him tremble, and we heard an interruption in a loud and clear voice; but without shouting, or clamour. What had been said? It seemed that it was: 'Down with the goddesses!' But I thought that I had not heard aright, so out of place and void of sense it seemed. However, the exclamation was repeated twice or thrice; then some one cried: 'Save the Archbishop!' Other voices replied: 'To arms!' The crowd, overturning the

chairs and the barriers, scattered, and rushed towards the doors shrieking. Amidst the wails of the children, and the screams of the women,
Éliphas, carried away by the crowd, found himself, somehow or other,
out of the church; but the last look that he was able to cast upon it was
smitten with a terrible and ineffaceable picture!

In the midst of a circle made large by the affright of all those who
surrounded him, the prelate was standing alone, leaning always on his
cross, and held up by the stiffness of his cope, which the grand-vicars
had let go, and which accordingly hung down to the ground.

The head of the Archbishop was a little thrown back, his eyes and
his free hand raised to Heaven. His attitude was that which Eugène
Delacroix has given to the Bishop of Liège in the picture of his assassination by the bandits of the Wild Boar of the Ardennes;[1] there was in
his gesture the whole epic of martyrdom; it was an acceptance and an
offering; a prayer for his people, and a pardon for his murderer.

The day was falling, and the church was beginning to grow dark.
The Archbishop, his arms raised to Heaven, lighted by a last ray which
penetrated the casements of the nave, stood out upon a dark background, where one could scarcely distinguish a pedestal without a
statue, on which were written these two words of the Passion of Christ:
ECCE HOMO! and farther in the background, an apocalyptic painting
representing the four plagues ready to let themselves loose upon the
world, and the whirlwinds of hell, following the dusty traces of the
pale horse of death.

Before the Archbishop, a lifted arm, sketched in shadow like an
infernal silhouette, held and brandished a knife. Policemen, sword in
hand, were running up.

[1] Extract from Sir Walter Scott's *Notes on the Murder of the Bishop of Liège*: 'The
Bishop's murder did not take place till 1482. In the months of August and September of
that year, *William de la Marck*, called 'The Wild Boar of the Ardennes', entered into a
conspiracy with the discontented citizens of Liège against their Bishop, Louis of Bourbon,
being aided with considerable sums of money by the King of France. By this means and
with the assistance of many murderers and banditti, who thronged to him as to a
leader befitting them, De la Marck assembled a body of troops. With this little army he
approached the city of Liège. Upon this, the citizens, who were engaged in the conspiracy, came to their Bishop, and, offering to stand by him to the death, exhorted him to
march out against these robbers. The Bishop, therefore, put himself at the head of a few
troops of his own, trusting to the assistance of the people of Liège. But as soon as they
came in sight of the enemy, the citizens, as before agreed, fled from the Bishop's banner,
and he was left with his own handful of adherents. At this moment De la Marck charged
at the head of his men with the expected success. The Bishop was brought before De la
Marck, who first cut him over the face, then murdered him with his own hand, and
caused his body to be exposed naked in the great square of Liège before St. Lambert's
Cathedral.'

Three years after the Bishop's death, Maximilian, Emperor of Austria, caused De la
Marck to be arrested at Utrecht, where he was beheaded in 1485.

And while all this tumult was going on at the bottom of the church, the singing of the litanies continued in the choir, as the harmony of the orbs of Heaven goes on for ever, careless of our revolutions and of our anguish.

Éliphas Lévi had been swept out of the church by the crowd. He had come out by the right-hand door. Almost at the same moment the left-hand door was flung violently open, and a furious group of men rushed out of the church.

This group was whirling around a man whom fifty arms seemed to hold, whom a hundred shaken fists sought to strike.

This man later complained of having been roughly handled by the police, but, as far as one could see in such an uproar, the police were rather protecting him against the exasperation of the mob.

Women were running after him shrieking: 'Kill him!'

'But what has he done?' cried other voices.

'The wretch! He has struck the Archbishop with his fist!' said the women.

Then others came out of the church, and contradictory accounts were flying to and fro.

'The Archbishop was frightened, and has fainted,' said some.

'He is dead!' replied others.

'Did you see the knife?' added a third comer. 'It is as long as a sabre, and the blood was streaming on the blade.'

'The poor Archbishop has lost one of his slippers,' remarked an old woman, joining her hands.

'It is nothing! It is nothing!' cried a woman who rented chairs. 'You can come back to the church: Monseigneur is not hurt; they have just said so from the pulpit.'

The crowd then made a movement to return to the church.

'Go! Go!' said at that very moment the grave and anguished voice of a priest. 'The office cannot be continued; we are going to close the church: it is profaned.'

'How is the Archbishop?' said a man.

'Sir,' replied the priest, 'the Archbishop is dying; perhaps even at this very moment he is dead!'

The crowd dispersed in consternation to spread the mournful news over Paris.

A bizarre incident happened to Éliphas, and made a kind of diversion for his deep sorrow at what had just passed.

At the moment of the uproar, an aged woman of the most respectable appearance had taken his arm, and claimed his protection.

He made it a duty to reply to this appeal, and when he had got out of the crowd with this lady: 'How happy I am,' said she, 'to have met a man who weeps for this great crime, for which, at this moment, so many wretches rejoice!'

'What are you saying, madam? How is it possible that there should exist beings so depraved as to rejoice at so great a misfortune?'

'Silence!' said the old lady; 'perhaps we are overheard. . . . Yes,' she added, lowering her voice; 'there are people who are exceedingly pleased at what has happened. And look there, just now, there was a man of sinister mien, who said to the anxious crowd, when they asked him what had happened: "Oh, it is nothing! It is a spider which has fallen." '[1]

'No, madam, you must have misunderstood. The crowd would not have suffered so abominable a remark, and the man would have been immediately arrested.'

'Would to God that all the world thought as you do!' said the lady.

Then she added: 'I recommend myself to your prayers, for I see clearly that you are a man of God.'

'Perhaps every one does not think so,' replied Éliphas.

'And what does the world matter to us?' replied the lady with vivacity; 'the world lies and calumniates, and is impious! It speaks evil of you, perhaps. I am not surprised at it, and if you knew what it says of me, you would easily understand why I despise its opinion!'

'The world speaks evil of you, madam?'

'Yes, in truth, and the greatest evil that can be said.'

'How so?'

'It accuses me of sacrilege.'

'You frighten me. Of what sacrilege, if you please?'

'Of an unworthy comedy that I am supposed to have played in order to deceive two children, on the mountain of the Salette.'

'What! You must be——'

'I am Mademoiselle de la Merlière.'

'I have heard speak of your trial, mademoiselle, and of the scandal which it caused, but it seems to me that your age and your position ought to have sheltered you from such an accusation.'

---

[1] This man was presumably Lévi himself. As 'the abominable authors of the *Grimoires* concealed 'child' beneath 'kid', so Lévi is careful to disguise his true attitude to the Church which he wished to destroy.—A.C.

'Come and see me, sir, and I will present you to my lawyer, M. Favre, who is a man of talent whom I wish to gain to God.'

Thus talking, the two companions had arrived at the Rue du Vieux Colombier. The lady thanked her improvised cavalier, and renewed her invitation to come to see her.

'I will try to do so,' said Éliphas; 'but if I come shall I ask the porter for Mlle de la Merlière?'

'Do not do so,' said she; 'I am not known under that name; ask for Mme Dutruck.'

'Dutruck, certainly, madam; I present my humble compliments.' And they separated.

The trial of the assassin began, and Éliphas, reading in the news-papers that the man was a priest, that he had belonged to the clergy of St. Germain l'Auxerrois, that he had been a country vicar, and that he seemed exalted to the point of madness, recalled the pale priest who, a year earlier, had been looking for the *grimoire* of Honorius. But the description which the public sheets gave of the criminal disagreed with the recollection of the Professor of Magic. In fact, the majority of the papers said that he had black hair. . . . 'It is not he, then,' thought Éliphas. 'However, I still keep in my ear and in my memory the words which would now be explained for me by this great crime: "You will soon learn something. Before long, you will hear me spoken of." '

The trial took place with all the frightful vicissitudes with which every one is familiar, and the accused was condemned to death.

The next day, Éliphas read in a legal newspaper the account of this unheard-of scene in the annals of justice, but a cloud passed over his eyes when he came to the description of the accused: 'He is blond.'

'It must be he,' said the Professor of Magic.

Some days afterwards, a person who had been able to sketch the convict during the trial, showed it to Éliphas.

'Let me copy this drawing,' said he, all trembling with fear.

He made the copy, and took it to his friend Desbarrolles, of whom he asked, without other explanation:

'Do you know this head?'

'Yes,' said Desbarrolles energetically. 'Wait a moment: yes, it is the mysterious priest whom we saw at Mme A——'s, and who wanted to make magical evocations.'

'Oh, well, my friend, you confirm me in my sad conviction. The man we saw, we shall never see again; the hand which you examined has become a bloody hand. We have heard speak of him, as he told us we should; that pale priest, do you know what was his name?'

'Oh, my God!' said Desbarrolles, changing colour, 'I am afraid to know it!'

'Well, you know it: it was the wretch Louis Verger!'

Some weeks after what we have just recorded, Éliphas Lévi was talking with a bookseller whose speciality was to make a collection of old books concerning the occult sciences. They were talking of the *grimoire* of Honorius.

'Nowadays, it is impossible to find it,' said the merchant. 'The last that I had in my hands I sold to a priest for a hundred francs.'

'A young priest? And do you remember what he looked like?'

'Oh, perfectly, but you ought to know him well yourself, for he told me he had seen you, and it is I who sent him to you.'

No more doubt, then; the unhappy priest had found the fatal *grimoire*, he had done the evocation, and prepared himself for the murder by a series of sacrileges. For this is in what the infernal evocations consist, according to the *grimoire* of Honorius:

'Choose a black cock, and give him the name of the spirit of darkness which one wishes to evoke.

'Kill the cock, and keep its heart, its tongue, and the first feather of its left wing.

'Dry the tongue and the heart, and reduce them to powder.

'Eat no meat and drink no wine, that day.

'On Tuesday, at dawn, say a mass of the angels.

'Trace upon the altar itself, with the feather of the cock dipped in the consecrated wine, certain diabolical signatures (those of Mr. Home's pencil, and the bloody hosts of Vintras).

'On Wednesday, prepare a taper of yellow wax; rise at midnight, and alone, in the church, begin the office of the dead.

'Mingle with this office infernal evocations.

'Finish the office by the light of a single taper, extinguish it immediately, and remain without light in the church thus profaned until sunrise.

'On Thursday, mingle with the consecrated water the powder of the tongue and heart of the black cock, and let the whole be swallowed by a male lamb of nine days old. . . .'

The hand refuses to write the rest. It is a mixture of brutalizing practices and revolting crimes, so constituted as to kill for evermore judgment and conscience.[1]

But in order to communicate with the phantom of absolute evil, to realize that phantom to the point of seeing and touching it, is it not necessary to be without conscience and without judgment?

There is doubtless the secret of this incredible perversity, of this murderous fury, of this unwholesome hate against all order, all ministry, all hierarchy, of this fury, above all, against the dogma which sanctifies peace, obedience, gentleness, purity, under so touching an emblem as that of a mother.

This wretch thought himself sure not to die. The Emperor, thought he, would be obliged to pardon him; an honourable exile awaited him; his crime would give him an enormous celebrity; his reveries would be bought for their weight in gold by the booksellers. He would become immensely rich, attract the notice of a great lady, and marry beyond the seas. It is by such promises that the phantom of the devil, long ago, lured Gilles de Laval, Seigneur of Retz, and made him wade from crime to crime. A man capable of evoking the devil, according to the rites of the *grimoire* of Honorius, has gone so far upon the road of evil that he is disposed to all kinds of hallucinations, and all lies. So, Verger slept in blood, to dream of I know not what abominable pantheon; and he awoke upon the scaffold.

But the aberrations of perversity do not constitute an insanity; the execution of this wretch proved it.

One knows what desperate resistance he made to his executioners. 'It is treason,' said he; 'I cannot die so! Only one hour, an hour to write to the Emperor! The Emperor is bound to save me.'

Who, then, was betraying him?

Who, then, had promised him life?

Who, then, had assured him beforehand of a clemency which was impossible, because it would revolt the conscience of the public?

Ask all that of the *grimoire* of Honorius!

Two incidents in this tragic story bear upon the phenomena produced by Mr. Home: the noise of the storm heard by the wicked priest in his early evocations, and the difficulty which he found in expressing his real thought in the presence of Éliphas Lévi.

One may also comment upon the apparition of the sinister man taking pleasure in the public grief, and uttering an indeed infernal word

[1] The great painter, dipping his brush in earthquake and eclipse, employs an excess of yellow.—A.C.

in the midst of the consternation of the crowd, an apparition only noticed by the ecstatic of La Salette, the too celebrated Mlle de la Merlière, who has the air after all of a worthy individual, but very excitable, and perhaps capable of acting and speaking without knowing it herself, under the influence of a sort of ascetic sleep-waking.

This word 'sleep-waking' brings us back to Mr. Home, and our anecdotes have not made us forget what the title of this work promised to our readers.

We ought, then, to tell them what Mr. Home is.

We keep our promise.

Mr. Home is an invalid suffering from a contagious sleep-waking.

This is an assertion.

It remains to us to give an explanation and a demonstration.

That explanation and demonstration, in order to be complete, demand a work sufficient to fill a book.

That book has been written, and we shall publish it shortly.

Here is the title:

The Reason of Miracles, or the Devil at the Tribunal of Science.[1]

'Why the devil?'

Because we have demonstrated by facts what Mr. de Mirville had, before us, incompletely set forth.

We say 'incompletely'; because the devil is, for Mr. de Mirville, a fantastic personage, while for us it is the misuse of a natural force.

A medium once said: 'Hell is not a place, it is a state.'

We shall be able to add: 'The devil is not a person or a force; it is a vice, and in consequence, a weakness.'

Let us return for a moment to the study of phenomena!

Mediums are, in general, of poor health and narrow limitations.

They can accomplish nothing extraordinary in the presence of calm and educated persons.

One must be accustomed to them before seeing or feeling anything.

The phenomena are not identical for all present. For example, where one will see a hand, another will perceive nothing but a whitish smoke.

Persons impressed by the magnetism of Mr. Home feel a sort of indisposition; it seems to them that the room turns round, and the temperature seems to them to grow rapidly lower.

---

[1] That was the title which we intended at that time to give to the book which we now publish.—E.L.

The miracles are more successful in the presence of a few people chosen by the medium himself.

In a meeting of several persons, it may be that all will see the miracles—with the exception of one, who will see absolutely nothing.

Among the persons who do see, all do not see the same thing.

Thus, for example:

One evening, at Mme de B——'s, the medium made appear a child which that lady had lost. Mme de B—— alone saw the child; Count de M—— saw a little whitish vapour, in the shape of a pyramid; the others saw nothing.

Everybody knows that certain substances, hashish, for example, intoxicate without taking away the use of reason, and cause to be seen with an astonishing vividness things which do not exist.

A great part of the phenomena of Mr. Home belong to a natural influence similar to that of hashish.

This is the reason why the medium refuses to operate except before a small number of persons chosen by himself.

The rest of these phenomena should be attributed to magnetic power.

To see anything at Mr. Home's *séances* is not a reassuring index of the health of him who sees.

And even if his health should be in other ways excellent, the vision indicates a transitory perturbation of the nervous apparatus in its relation to imagination and light.

If this perturbation were frequently repeated, he would become seriously ill.

Who knows how many collapses, attacks of tetanus, insanities, violent deaths, the mania of table-turning has already produced?

These phenomena become particularly terrible when perversity takes possession of them.

It is then that one can really affirm the intervention and the presence of the spirit of evil.

Perversity or fatality, these pretended miracles obey one of these two powers.

As to qabalistic writings and mysterious signatures, we shall say that they reproduce themselves by the magnetic intuition of the mirages of thought in the universal vital fluid.

These instinctive reflections may be produced if the magic Word has nothing arbitrary in it, and if the signs of the occult sanctuary are the natural expressions of absolute ideas.

It is this which we shall demonstrate in our book.

But, in order not to send back our readers from the unknown to the future, we shall detach beforehand two chapters of that unpublished work, one upon the Qabalistic Word, the other upon the secrets of the Qabalah, and we shall draw conclusions which will complete in a manner satisfactory to all the explanation which we have promised in the matter of Mr. Home.

There exists a power which generates forms; this power is light.

Light creates forms in accordance with the laws of eternal mathematics, by the universal equilibrium of light and shadow.

The primitive signs of thought trace themselves by themselves in the light, which is the material instrument of thought.

God is the soul of light. The universal and infinite light is for us, as it were, the body of God.

The Qabalah, or transcendental magic, is the science of light.

Light corresponds to life.

The kingdom of shadows is death.

All the dogmas of true religion are written in the Qabalah in characters of light upon a page of shadow.

The page of shadows consists of blind beliefs.

Light is the great plastic medium.

The alliance of the soul and the body is a marriage of light and shadow.

Light is the instrument of the Word, it is the white writing of God upon the great book of night.

Light is the source of thought, and it is in it that one must seek for the origin of all religious dogma. But there is only one true dogma, as there is only one pure light; shadow alone is infinitely varied.

Light, shadow, and their harmony, which is the vision of beings, form the principle analogous to the great dogmas of Trinity, of Incarnation, and of Redemption.

Such is also the mystery of the Cross.

It will be easy for us to prove this by an appeal to religious monu ments, by the signs of the primitive Word, by those books which contain the secrets of the Qabalah, and finally by the reasoned explanation of all the mysteries by the means of the keys of qabalistic magic.

In all symbolisms, in fact, we find ideas of antagonism and of harmony producing a trinitarian notion in the conception of divinity, following which the mythological personification of the four cardinal points of Heaven completes the sacred septenary, the base of all dogmas and of all rites. In order to convince oneself of it, it is sufficient to read again and meditate upon the learned work of Dupuis, who would be a great qabalist if he had seen a harmony of truths where his negative preoccupations only permitted him to see a concert of errors.

It is not here our business to repeat his work, which everybody knows; but it is important to prove that the religious reform brought about by Moses was altogether qabalistic, that Christianity, in instituting a new dogma, has simply come nearer to the primitive sources of the teachings of Moses, and that the Gospel is no more than a transparent veil thrown upon the universal and natural mysteries of oriental initiation.

A distinguished but little-known man of learning, Mr. P. Lacour, in his book on the Elohim or Mosaic God, has thrown a great light on that question, and has rediscovered in the symbols of Egypt all the allegorical figures of Genesis. More recently, another courageous student of vast erudition, Mr. Vincent (de l'Yonne), has published a treatise upon idolatry among both the ancients and the moderns, in which he raises the veil of universal mythology.

We invite conscientious students to read these various works, and we confine ourselves to the special study of the Qabalah among the Hebrews.

The Logos, or the Word, being according to the initiates of that science the complete revelation, the principles of the holy Qabalah ought to be found reunited in the signs themselves of which the primitive alphabet is composed.

Now, this is what we find in all Hebrew grammars.[1]

There is a fundamental and universal letter which generates all the others. It is the IOD.

There are two other mother letters, opposed and analogous among themselves; the ALEPH א and the MEM מ, according to others, the SHIN ש.

There are seven double letters, the BETH ב, the GIMEL ג, the DALETH ד, the KAPH כ, the PE פ, the RESH ר, and the TAU ת.

Finally, there are twelve simple letters; in all twenty-two. The unity is represented, in a relative manner, by the ALEPH; the ternary is figured either by IOD, MEM, SHIN, or by ALEPH, MEM, SHIN.

The septenary, by BETH, GIMEL, DALETH, KAPH, PE, RESH, TAU.

---

[1] This is all deliberately wrong. That Lévi knew the correct attributions is evident from a MS. annotated by himself. Lévi refused to reveal these attributions, rightly enough, as his grade was not high enough, and the time not ripe. Note the subtlety of the form of his statement. The correct attributions are in Liber 777.—A.C.

The duodenary is the ternary multiplied by four; and it re-enters thus into the symbolism of the septenary.

Each letter represents a number: each assemblage of letters, a series of numbers.

The numbers represent absolute philosophical ideas.

The letters are shorthand hieroglyphs.

Let us see now the hieroglyphic and philosophical significations of each of the twenty-two letters (*vide* Bellarmin, Reuchlin, Saint-Jérôme, Kabala Denudata, Sepher Yetzirah, Technica Curiosa of Father Schott, Picus de Mirandola, and other authors, especially those of the collection of Pistorius).

The Mothers

The IOD. The absolute principle, the productive being.

The MEM. Spirit, or the Jakin of Solomon.

The SHIN. Matter, or the column called Boaz.

The Double Letters

BETH. Reflection, thought, the moon, the Angel Gabriel, Prince of mysteries.

GIMEL. Love, will, Venus, the Angel Anael, Prince of life and death.

DALETH. Force, power, Jupiter, Sachiel, Melech, King of kings.

KAPH. Violence, strife, work, Mars, Samael Zebaoth, Prince of Phalanges.

PE. Eloquence, intelligence, Mercury, Raphael, Prince of sciences.

RESH. Destruction and regeneration, Time, Saturn, Cassiel, King of tombs and of solitude.

TAU. Truth, light, the Sun, Michael, King of the Elohim.

The Simple Letters

The simple letters are divided into four triplicities, having for titles the four letters of the divine tetragram יהוה.

In the divine tetragram, the IOD, as we have just said, symbolizes the productive and active principle. The HÉ ה represents the passive productive principle, the CTEIS. The VAU symbolizes the union of the two, or the lingam, and the final HÉ is the image of the second reproductive principle; that is to say, of the passive reproduction in the world of effects and forms.

The twelve simple letters, הוזחטלנסעצק and י or מ, divided into threes, reproduce the notion of the primitive triangle, with the interpretation, and under the influence, of each of the letters of the tetragram.

One sees that the philosophy and the religious dogma of the Qabalah are there indicated in a complete but veiled manner.

Let us now investigate the allegories of Genesis.

'In the beginning (IOD the unity of being,) Elohim, the equilibrated forces (Jakin and Boaz), created the Heaven (spirit) and the earth (matter), or in other words, good and evil, affirmation and negation.' Thus begins the Mosaic account of creation.

Then, when it comes to giving a place to man, and a sanctuary to his alliance with divinity, Moses speaks of a garden, in the midst of which a single fountain branched into four rivers (the IOD and the TETRAGRAM), and then of two trees, one of life, and the other of death, planted near the river. There are placed the man and the woman, the active and the passive; the woman sympathizes with death, and draws Adam with her in her fall. They are then driven out from the sanctuary of truth, and a kerub (a bull-headed sphinx, *vide* the hieroglyphs of Assyria, of India and of Egypt) is placed at the gate of the garden of truth in order to prevent the profane from destroying the tree of life. Here we have mysterious dogma, with all its allegories and its terrors, replacing the simplicity of truth. The idol has replaced God, and fallen humanity will not delay to give itself up to the worship of the golden calf.

The mystery of the necessary and successive reactions of the two principles on each other is indicated subsequently by the allegory of Cain and Abel. Force avenges itself by oppression for the seduction of weakness; martyred weakness expiates and intercedes for force when it is condemned for its crime to branding remorse. Thus is revealed the equilibrium of the moral world; here is the basis of all the prophecies, and the fulcrum of all intelligent political thought. To abandon a force to its own excesses is to condemn it to suicide.

Dupuis failed to understand the universal religious dogma of the Qabalah, because he had not the science of the beautiful hypothesis, partly demonstrated and realized more from day to day by the discoveries of science: I refer to *universal analogy*.

Deprived of this key of transcendental dogma, he could see no more of the gods than the sun, the seven planets, and the twelve signs of the zodiac; but he did not see in the sun the image of the Logos of Plato, in the seven planets the seven notes of the celestial gamut, and in the zodiac the quadrature of the ternary circle of all initiations.

The Emperor Julian, that *adept of the spirit* who was never understood, that initiate whose paganism was less idolatrous than the faith of certain Christians, the Emperor Julian, we say, understood better than Dupuis and Volney the symbolic worship of the sun. In his hymn to the king, Helios, he recognizes that the star of day is but the reflection and the material shadow of that sun of truth which illumines the world of intelligence, and which is itself only a light borrowed from the Absolute.

It is a remarkable thing that Julian has ideas of the Supreme God, that the Christians thought they alone adored, much greater and more correct than those of some of the fathers of the Church, who were his contemporaries, and his adversaries.

This is how he expresses himself in his defence of Hellenism:

'It is not sufficient to write in a book that God spake, and things were made. It is necessary to examine whether the things that one attributes to God are not contrary to the very laws of Being. For, if it is so, God could not have made them, for He could not contradict Nature without denying Himself. . . . God being eternal, it is of the nature of necessity that His orders should be immutable as He.'

So spake that apostate, that man of impiety! Yet, later, a Christian doctor, become the oracle of the theological schools, taking his inspiration perhaps from these splendid words of the misbeliever, found himself obliged to bridle superstition by writing that beautiful and brave maxim which easily resumes the thought of the great Emperor:

'A thing is not just because God wills it; but God wills it because it is just.'

The idea of a perfect and immutable order in nature, the notion of an ascending hierarchy and of a descending influence in all beings, had furnished to the ancient hierophants the first classification of the whole of natural history. Minerals, vegetables, animals were studied analogically; and they attributed their origin and their properties to the passive or to the active principle, to the darkness or to the light. The sign of their election or of their reprobation, traced in their natural form, became the hieroglyphic character of a vice or a virtue; then, by dint of taking the sign for the thing, and expressing the thing by the sign, they ended by confounding them. Such is the origin of that fabulous natural history, in which lions allow themselves to be defeated by cocks, where dolphins die of sorrow for the ingratitude of men, in which mandrakes speak, and the stars sing. This enchanted world is indeed the poetic domain of magic; but it has no other reality than the meaning of the hieroglyphs which gave it birth. For the sage who understands the analogies of the transcendental Qabalah, and the exact relation of ideas with signs, this fabulous country of the fairies is a country still fertile in discoveries; for those truths which are too beautiful, or too simple to please men, without any veil, have all been hidden in these ingenious shadows.

Yes, the cock can intimidate the lion, and make himself master of him, because vigilance often supplants force, and succeeds in taming wrath. The other fables of the sham natural history of the ancients are explained in the same manner, and in this allegorical use of analogies, one can already understand the possible abuses and predict the errors to which the Qabalah was obliged to give birth.

The law of analogies, in fact, has been for qabalists of a secondary rank the object of a blind and fanatical faith. It is to this belief that one must attribute all the superstitions with which the adepts of occult science have been reproached. This is how they reasoned:

The sign expresses the thing.

The thing is the virtue of the sign.

There is an analogical correspondence between the sign and the thing signified.

The more perfect is the sign, the more entire is the correspondence.

To say a word is to evoke a thought and make it present. To name God is to manifest God.

The word acts upon souls, and souls react upon bodies; consequently one can frighten, console, cause to fall ill, cure, even kill, and raise from the dead by means of words.

To utter a name is to create or evoke a being.

In the name is contained the *verbal* or spiritual doctrine of the being itself.

When the soul evokes a thought, the sign of that thought is written automatically in the light.

To invoke is to adjure, that is to say, to swear by a name; it is to perform an act of faith in that name, and to communicate in the virtue which it represents.

Words in themselves are, then, good or evil, poisonous or wholesome.

The most dangerous words are vain and lightly uttered words, because they are the voluntary abortions of thought.

A useless word is a crime against the spirit of intelligence; it is an intellectual infanticide.

Things are for every one what he makes of them by naming them. The *word* of every one is an impression or an habitual prayer.

To speak well is to live well.

A fine style is an aureole of holiness.

From these principles, some true, others hypothetical, and from the more or less exaggerated consequences that they draw from them, there resulted for superstitious qabalists an absolute confidence in enchantments, evocations, conjurations and mysterious prayers. Now, as faith has always accomplished miracles, apparitions, oracles, mysterious cures, sudden and strange maladies, have never been lacking to it.

It is thus that a simple and sublime philosophy has become the secret science of Black Magic. It is from this point of view above all that the Qabalah is still able to excite the curiosity of the majority in our so distrustful and so credulous century. However, as we have just explained, that is not the true science.

Men rarely seek the truth for its own sake; they have always a secret motive in their efforts, some passion to satisfy, or some greed to assuage. Among the secrets of the Qabalah there is one above all which has always tormented seekers; it is the secret of the transmutation of metals, and of the conversion of all earthly substances into gold.

Alchemy borrowed all these signs from the Qabalah, and it is upon the law of the analogies resulting from the harmony of contraries that it based its operations. An immense physical secret was, moreover, hidden under the qabalistic parables of the ancients. This secret we have arrived at deciphering, and we shall submit its letter to the investigations of the gold-makers. Here it is:

1°. The four imponderable fluids are nothing but the diverse manifestations of one same universal agent, which is light.

2°. Light is the fire which serves for the Great Work under the form of electricity.

3°. The human will directs the vital light by means of the nervous system.

4°. The secret agent of the Great Work, the Azoth of the sages, the living and life-giving gold of the philosophers, the universal metallic productive agent, is MAGNETIZED ELECTRICITY.[1]

The alliance of these two words still does not tell us much, and yet, perhaps, they contain a force sufficient to overturn the world. We say 'perhaps' on philosophical grounds, for, personally, we have no doubt whatever of the high importance of this great hermetic arcanum.

We have just said that alchemy is the daughter of the Qabalah; to convince oneself of the truth of this it is sufficient to look at the symbols of Flamel, of Basil Valentine, the pages of the Jew Abraham, and the more or less apocryphal oracles of the Emerald Table of Hermes. Everywhere one finds the traces of that decade of Pythagoras, which is so magnificently applied in the Sepher Yetzirah to the complete and absolute notion of divine things, that decade composed of unity and triple ternary which the Rabbis have called the Berashith, and the Mercavah, the luminous tree of the Sephiroth, and the key of the Shemhamphorash.

We have spoken at some length in our book entitled *Transcendental Magic* of a hieroglyphic monument (preserved up to our own time under a futile pretext) which alone explains all the mysterious writings of high initiation. This monument is that Tarot of the Bohemians which gave rise to our games of cards. It is composed of twenty-two allegorical letters, and of four series of ten hieroglyphs each, referring to the four letters of the name of Jehovah. The diverse combinations of those signs, and the numbers which correspond to them, form so many qabalistic oracles, so that the whole science is contained in this mysterious book. This perfectly simple philosophical machine astonishes by the depth of its results.

The Abbé Trithemius, one of our greatest masters in magic, composed a very ingenious work, which he calls Polygraphy, upon the qabalistic alphabet. It is a combined series of progressive alphabets where each letter represents a word, the words correspond to each other, and complete themselves from one alphabet to another; and there is no doubt that Trithemius was acquainted with the Tarot, and made use of it to set his learned combinations in logical order.

Jerome Cardan was acquainted with the symbolical alphabet of the initiates, as one may recognize by the number and disposition of the chapters of his work on Subtlety. This work, in fact, is composed of

---

[1] In this joke, Lévi indicates that he really knew the Great Arcanum; but only those who also possess it can recognize it, and enjoy the joke.—A.C.

twenty-two chapters, and the subject of each chapter is analogous to the number and to the allegory of the corresponding card of the Tarot. We have made the same observation on a book of St. Martin entitled *A Natural Picture of the Relations which exist between God, Man and the Universe.* The tradition of this secret has, then, never been interrupted from the first ages of the Qabalah to our own times.

The table-turners, and those who make the spirits speak with alphabetical charts, are, then, a good many centuries behind the times; they do not know that there exists an oracular instrument whose words are always clear and always accurate, by means of which one can communicate with the seven genii of the planets, and make to speak at will the seventy-two wheels of Assiah, of Yetzirah, and of Briah. For that purpose it is sufficient to understand the system of universal analogies, such as Swedenborg has set forth in the hieroglyphic key of the arcana; then to mix the cards together, and draw from them by chance, always grouping them by the numbers corresponding to the ideas on which one desires enlightenment; then, reading the oracles as qabalistic writings ought to be read, that is to say, beginning in the middle and going from right to left for odd numbers, beginning on the right for even numbers, and interpreting successively the number for the letter which corresponds to it, the grouping of the letters by the addition of their numbers, and all the successive oracles by their numerical order, and their hieroglyphic relations.

This operation of the qabalistic sages, originally intended to discover the rigorous development of absolute ideas, degenerated into superstition when it fell into the hands of the ignorant priests and the nomadic ancestors of the Bohemians who possessed the Tarot in the Middle Ages; they did not know how to employ it properly, and used it solely for fortune-telling.

The game of chess, attributed to Palamedes, has no other origin than the Tarot, and one finds there the same combinations and the same symbols; the king, the queen, the knight, the soldier, the fool, the tower, and houses representing numbers. In old times, chess-players sought upon their chess-board the solution of philosophical and religious problems, and argued silently with each other in manœuvring the hieroglyphic characters across the numbers. Our vulgar game of goose, revived from the old Grecian game, and also attributed to Palamedes, is nothing but a chess-board with motionless figures and numbers movable by means of dice. It is a Tarot disposed in the form of a wheel, for the use of aspirants to initiation. Now, the word Tarot, in which one finds 'rota' and 'tora', itself expresses, as William Postel has demonstrated, this primitive disposition in the form of a wheel.

The hieroglyphs of the game of goose are simpler than those of the Tarot, but one finds the same symbols in it: the juggler, the king, the queen, the tower, the devil or Typhon, death, and so on. The dice-indicated chances of the game represent those of life, and conceal a highly philosophical sense sufficiently profound to make sages meditate, and simple enough to be understood by children.

The allegorical personage Palamedes is, however, identical with Enoch, Hermes, and Cadmus, to whom various mythologies have attributed the invention of letters. But, in the conception of Homer, Palamedes, the man who exposed the fraud of Ulysses and fell a victim to his revenge, represents the initiator or the man of genius whose eternal destiny is to be killed by those whom he initiates. The disciple does not become the living realization of the thoughts of the Master until he has drunk his blood and eaten his flesh, to use the energetic and allegorical expression of the initiator, so ill understood by Christians.

The conception of the primitive alphabet was, as one may easily see, the idea of a universal language which should enclose in its combinations, and even in its signs themselves, the recapitulation and the evolutionary law of all sciences, divine and human. In our own opinion, nothing finer or greater has ever been dreamt by the genius of man; and we are convinced that the discovery of this secret of the ancient world has fully repaid us for so many years of sterile research and thankless toil in the crypts of lost sciences and the cemeteries of the past.

. One of the first results of this discovery should be to give a new direction to the study of the hieroglyphic writings as yet so imperfectly deciphered by the rivals and successors of M. Champollion.

The system of writing of the disciples of Hermes being analogical and synthetical, like all the signs of the Qabalah, would it not be useful, in order to read the pages engraved upon the stones of the ancient temples, to replace these stones in their place, and to count the numbers of their letters, comparing them with the numbers of other stones?

The obelisk of Luxor, for example, was it not one of the two columns at the entrance of a temple? Was it at the right-hand or the left-hand pillar? If at the right, these signs refer to the active principle; if at the left, it is by the passive principle that one must interpret its characters. But there should be an exact correspondence of one obelisk with the other, and each sign should receive its complete sense from the analogy of contraries. M. Champollion found Coptic in the hieroglyphics, another savant would perhaps find more easily, and more fortunately, Hebrew; but what would one say if it were neither Hebrew nor Coptic? If it were, for example, the universal primitive language? Now, this language, which was that of the transcendental Qabalah, did certainly exist; more, it still exists at the base of Hebrew itself, and of all the oriental languages which derive from it; this language is that of the sanctuary, and the columns at the entrance of the temples ordinarily contained all its symbols. The intuition of the ecstatics comes nearer to the truth with regard to these primitive signs than even the science of the learned, because, as we have said, the universal vital fluid, the astral light, being the mediating principle between the ideas and the forms, is obedient to the extraordinary leaps of the soul which seeks the unknown, and furnishes it naturally with the signs already found, but forgotten, of the great revelations of occultism. Thus are formed the pretended signatures of spirits, thus were produced the mysterious writings of Gablidone who appeared to Dr. Lavater, the phantoms of Schroepfer, of St. Michel-Vintras, and the spirits of Mr. Home.

If electricity can move a light, or even a heavy, body without one touching it, is it impossible to give by magnetism a direction to electricity, and to produce, thus naturally, signs and writings? One can do it, doubtless; because one does it.

Thus, then, to those who ask us: 'What is the most important agent of miracles?' we shall reply—

'It is the first matter of the Great Work.'

'IT IS MAGNETIZED ELECTRICITY.'

Everything has been created by light.

It is in light that form is preserved.

It is by light that form reproduces itself.

The vibrations of light are the principle of universal movement.

By light, the suns are attached to each other, and they interlace their rays like chains of electricity.

Men and things are magnetized by light like the suns, and, by means of electro-magnetic chains whose tension is caused by sympathies and affinities, are able to communicate with each other from one end of the world to the other, to caress or strike, wound or heal, in a manner doubtless natural, but invisible, and of the nature of prodigy.

There is the secret of magic.

Magic, that science which comes to us from the magi!

Magic, the first of sciences!

Magic, the holiest science, because it establishes in the sublimest manner the great religious truths!

Magic, the most calumniated of all, because the vulgar obstinately confound magic with the superstitious sorcery whose abominable practices we have denounced!

It is only by magic that one can reply to the enigmatical questions of the Sphinx of Thebes, and find the solution of those problems of religious history which are sealed in the sometimes scandalous obscurities which are to be found in the stories of the Bible.

The sacred historians themselves recognize the existence and the power of the magic which boldly rivalled that of Moses.

The Bible tells us that Jannes and Jambres, Pharaoh's magicians, at first performed *the same miracles* as Moses, and that they declared those which they could not imitate impossible to human science. It is in fact more flattering to the self-love of a charlatan to deem that a miracle has taken place, than to declare himself conquered by the science or skill of a fellow-magician—above all, when he is a political enemy or a religious adversary.

When does the possible in magical miracles begin and end? Here is a serious and important question. What is certain is the existence of the facts which one habitually describes as miracles. Magnetizers and sleep-wakers do them every day; Sister Rose Tamisier did them; the 'Illuminated' Vintras does them still; more than fifteen thousand witnesses recently attested those of the American mediums; ten thousand peasants of Berry and Sologne would attest, if need were, those of the god Cheneau (a retired button-merchant who believes himself inspired by God). Are all these people hallucinated or knaves? Hallucinated, yes, perhaps, but the very fact that their hallucination is identical, whether separately or collectively, is it not a sufficiently great miracle on the part of him who produces it, always, at will, and at a stated time and place?

To do miracles, and to persuade the multitude that one does them, are very nearly the same thing, above all in a century as frivolous and scoffing as ours. Now, the world is full of wonder-makers, and science is often reduced to denying their works or refusing to see them, in order not to be reduced to examining them, or assigning a cause to them.

In the last century all Europe resounded with the miracles of Cagliostro. Who is ignorant of what powers were attributed to his 'wine of Egypt', and to his "elixir'? What can we add to the stories that they tell of his other-world suppers, where he made appear in flesh and blood the illustrious personages of the past? Cagliostro was, however, far from being an initiate of the first order, since the Great White Brotherhood abandoned him[1] to the Roman Inquisition, before whom he made, if one can believe the documents of his trial, so ridiculous and so odious an explanation of the Masonic trigram, L∴ P∴ D∴

But miracles are not the exclusive privilege of the first order of initiates; they are often performed by beings without education or virtue. Natural laws find an opportunity in an organism whose exceptional qualifications are not clear to us, and they perform their work with their invariable precision and calm. The most refined gourmets appreciate truffles, and employ them for their purposes, but it is hogs that dig them up: it is analogically the same for plenty of things less material and less gastronomical: instincts have groping presentiments, but it is really only science which discovers.

The actual progress of human knowledge has diminished by a great deal the chances of prodigies, but there still remains a great number, since both the power of the imagination and the nature and power of magnetism are not yet known. The observation of universal analogies, moreover, has been neglected, and for that reason divination is no longer believed in.

A qabalistic sage may, then, still astonish the crowd and even bewilder the educated:

1°—by divining hidden things; 2°—by predicting many things to come; 3°—by dominating the will of others so as to prevent them doing what they will, and forcing them to do what they do not will; 4°—by exciting apparitions and dreams; 5°—by curing a large number of illnesses; 6°—by restoring life to subjects who display all the symptoms of death; 7°—lastly, by demonstrating (if need be, by examples) the reality of the philosophical stone, and the transmutation of metals, according to the secrets of Abraham the Jew, of Flamel, and of Raymond Lully.

All these prodigies are accomplished by means of a single agent which the Hebrew calls OD, as did the Chevalier de Reichenbach, which we, with the School of Pasqualis Martinez, call astral light, which Mr.

---

[1] This is no more an argument than to say that God 'abandoned' Christ. Martyrdom is usually cited on the other side. Besides, the fate of Cagliostro is unknown—at least to the world at large.—A.C.

de Mirville calls the devil, and which the ancient alchemists called Azoth. It is the vital element which manifests itself by the phenomena of heat, light, electricity and magnetism, which magnetizes all terrestrial globes, and all living beings.

In this agent even are manifested the proofs of the qabalistic doctrine with regard to equilibrium and motion, by double polarity; when one pole attracts the other repels, one produces heat, the other cold, one gives a blue or greenish light, the other a yellow or reddish light.

This agent, by its different methods of magnetization, attracts us to each other, or estranges us from each other, subordinates one to the wishes of the other by causing him to enter his centre of attraction, re-establishes or disturbs the equilibrium in animal economy by its transmutations and its alternate currents, receives and transmits the imprints of the force of imagination which is in men the image and the semblance of the creative word, and thus produces presentiments and determines dreams. The science of miracles is then the knowledge of this marvellous force, and the art of doing miracles is simply the art of magnetizing or *illuminating* beings, according to the invariable laws of magnetism or astral light.

We prefer the word 'light' to the word 'magnetism', because it is more traditional in occultism, and expresses in a more complete and perfect manner the nature of the secret agent. There is, in truth, the liquid and drinkable gold of the masters in alchemy; the word 'OR' (the French word for 'gold') comes from the Hebrew 'AOUR' which signifies 'light'. 'What do you wish?' they asked the candidate in every initiation: 'To see the light,' should be the answer. The name of illuminati, which one ordinarily gives to adepts, has then been generally very badly interpreted by giving to it a mystical sense, as if it signified men whose intelligence believes itself to be lighted by a miraculous day. 'Illuminati' means, simply, knowers and possessors of the light, either by the knowledge of the great magical agent, or by the rational and ontological notion of the absolute.

The universal agent is a force tractable and subordinate to intelligence. Abandoned to itself, it, like Moloch, devours rapidly all that to which it gives birth, and changes the superabundance of life into immense destruction. It is, then, the infernal serpent of the ancient myths, the Typhon of the Egyptians, and the Moloch of Phoenicia; but if Wisdom, mother of the Elohim, puts her foot upon his head, she outwears all the flames which he belches forth, and pours with full hands upon the earth a vivifying light. Thus also it is said in the Zohar that at the beginning of our earthly period, when the elements disputed among themselves the surface of the earth, that fire, like an immense serpent, had enveloped everything in its coils, and was about to consume all beings, when divine clemency, raising around it the waves of the sea like a vestment of clouds, put her foot upon the head of the serpent and made him re-enter the abyss. Who does not see in this allegory the first idea, and the most reasonable explanation, of one of the images dearest to Catholic symbolism, the triumph of the Mother of God?

The qabalists say that the occult name of the devil, his true name, is that of Jehovah written backwards. This, for the initiate, is a complete revelation of the mysteries of the tetragram. In fact, the order of the letters of that great name indicates the predominance of the idea over form, of the active over the passive, of cause over effect. By reversing that order one obtains the contrary. Jehovah is he who tames Nature as it were a superb horse and makes it go where he will; Chavajoh (the demon) is the horse without a bridle who, like those of the Egyptians of the song of Moses, falls upon its rider, and hurls him beneath it, into the abyss.

The devil, then, exists really enough for the qabalists; but it is neither a person nor a distinguished power of even the forces of Nature. The devil is dispersion, or the slumber of the intelligence. It is madness and falsehood.

Thus are explained the nightmares of the Middle Ages; thus, too, are explained the bizarre symbols of some initiates, those of the Templars, for example, who are much less to be blamed for having worshipped Baphomet, than for allowing its image to be perceived by the profane. Baphomet, pantheistic figure of the universal agent, is nothing else than the bearded devil of the alchemists. One knows that the members of the highest grades in the old hermetic masonry attributed to a bearded demon the accomplishment of the Great Work. At this word, the vulgar hastened to cross themselves, and to hide their eyes, but the initiates of the cult of Hermes-Pantheos understood the allegory, and were very careful not to explain it to the profane.

Mr. de Mirville, in a book today almost forgotten, though it made some noise a few months ago, gives himself a great deal of trouble to compile an account of various sorceries, of the kind which fill the compilations of people like Delancre, Delrio, and Bodin. He might have found better than that in history. And without speaking of the easily attested miracles of the Jansenists of Port Royal, and of the Deacon Paris, what is more marvellous than the great monomania of martyrdom which has made children, and even women, during three hundred years, go to execution as if to a feast? What more magnificent than that enthusiastic faith accorded during so many centuries to the most incomprehensible, and, humanly speaking, to the most revolting mysteries? On this occasion, you will say, the miracles came from God, and one even employs them as a proof of the truth of religion. But, what? Heretics, too, let themselves be killed for dogmas. They then sacrificed both their reason and their life to their belief? Oh, for heretics, it is evident that the devil was responsible. Poor folk, who took the devil for God, and God for the devil! Why have they not been undeceived by making them recognize the true God by the charity, the knowledge, the justice, and above all, by the mercy of His ministers?

The necromancers who cause the devil to appear after a fatiguing and almost impossible series of the most revolting evocations, are only children by the side of that St. Anthony of the legend who drew them from hell by thousands, and dragged them everywhere after him like Orpheus, who attracted to him oaks, rocks, and the most savage animals.

Callot alone, initiated by the wandering Bohemians during his infancy into the mysteries of black sorcery, was able to understand and reproduce the evocations of the first hermit. And do you think that in retracing those frightful dreams of maceration and fasting, the makers of legends have invented? No; they have remained far below the truth. The cloisters, in fact, have always been peopled with nameless spectres, and their walls have palpitated with shadows and infernal larvae. St. Catherine of Siena on one occasion passed a week in the midst of an obscene orgy which would have discouraged the lust of Pietro di Aretino; St. Theresa felt herself carried away living into hell, and there suffered, between walls which ever closed upon her, tortures which only hysterical women will be able to understand. . . . All that, one will say, happened in the imagination of the sufferers. But where, then, would you expect facts of a supernatural order to take place? What is certain is that all these visionaries have seen and touched, that they have had the most vivid feeling of a formidable reality. We speak of it from our own experience, and there are visions of our own first youth, passed in retreat and asceticism, whose memory makes us shudder even now.

God and the devil are the ideals of absolute good and evil. But man never conceives absolute evil, save as a false idea of good. Good only can be absolute; and evil is only relative to our ignorance, and to our errors. Every man, in order to be a god, first makes himself a devil; but as the law of solidarity is universal, the hierarchy exists in hell as it does in Heaven. A wicked man will always find one more wicked than himself to do him harm; and when the evil is at its climax, it must cease, for it could only continue by the annihilation of being, which is impossible. Then the man-devils, at the end of their resources, fall once more under the empire of the god-men, and are saved by those whom one at first thought their victims; but the man who strives to live a life of evil deeds, does homage to good by all the intelligence and energy that he develops in himself. For this reason the great initiator said in his figurative language: 'I would that thou wert cold or hot; but because thou art lukewarm, I will spew thee out of my mouth.'

The Great Master, in one of His parables, condemns only the idle man who buried his treasure from fear of losing it in the risky operations of that bank which we call life. To think nothing, to love nothing, to wish for nothing, to do nothing—that is the real sin. Nature only recognizes and rewards workers.

The human will develops itself and increases itself by its own activity. In order to will truly, one must act. Action always dominates inertia and drags it at its chariot wheels. This is the secret of the influence of the alleged wicked over the alleged good. How many poltroons and cowards think themselves virtuous because they are afraid to be otherwise! How many respectable women cast an envious eye upon prostitutes! It is not very long ago since convicts were in fashion. Why? Do you think that public opinion can ever give homage to vice? No, but it can do justice to activity and bravery, and it is right that cowardly knaves should esteem bold brigands.

Boldness united to intelligence is the mother of all successes in this world. To undertake, one must know; to accomplish, one must will; to will really, one must dare; and in order to gather in peace the fruits of one's audacity, one must keep silent.

TO KNOW, TO DARE, TO WILL, TO KEEP SILENT, are, as we have said elsewhere, the four qabalistic words which correspond to the four letters of the tetragram and to the four hieroglyphic forms of the Sphinx. To know, is the human head; to dare, the claws of the lion; to will, the mighty flanks of the bull; to keep silent, the mystical wings of the eagle. He only maintains his position above other men who does not prostitute the secrets of his intelligence to their commentary and their laughter.

All men who are really strong are magnetizers, and the universal agent obeys their will. It is thus that they work marvels. They make themselves believed, they make themselves followed, and when they say: 'This is thus,' Nature changes (in a sense) to the eyes of the vulgar, and becomes what the great man wished. 'This is my flesh and this is my blood,' said a Man who had made himself God by His virtues; and eighteen centuries, in the presence of a piece of bread and a little wine, have seen, touched, tasted and adored flesh and blood made divine by martyrdom! Say now that the human will accomplished no miracles!

Do not let us here speak of Voltaire! Voltaire was not a wonder-worker, he was the witty and eloquent interpreter of those on whom the miracle no longer acted. Everything in his work is negative; everything was affirmative, on the contrary, in that of the 'Galilean', as an illustrious and too unfortunate Emperor called Him.

And yet Julian in his time attempted more than Voltaire could accomplish; he wished to oppose miracles to miracles, the austerity of power to that of revolt, virtues to virtues, wonders to wonders; the Christians never had a more dangerous enemy, and they recognized the fact, for Julian was assassinated; and the Golden Legend still bears witness that a holy martyr, awakened in his tomb by the clamour of the Church, resumed his arms, and struck the Apostate in the darkness, in the midst of his army and of his victories. Sorry martyrs, who rise from the dead to become hangmen! Too credulous Emperor, who believed in his gods, and in the virtues of the past!

When the kings of France were hedged around with the adoration of their people, when they were regarded as the Lord's anointed, and the eldest sons of the Church, they cured scrofula. A man who is the fashion can always do miracles when he wishes. Cagliostro may have been only a charlatan, but as soon as opinion had made of him 'the divine Cagliostro', he was expected to work miracles; and they happened.

When Cephas Barjona was nothing but a Jew proscribed by Nero, retailing to the wives of slaves a specific for eternal life, Cephas Barjona, for all educated people of Rome, was only a charlatan; but public opinion made an apostle of the spiritualistic empiric; and the successors of Peter, were they Alexander VI, or even John XXII, are infallible for every man who is properly brought up, who does not wish to put himself uselessly outside the pale of society. So goes the world.

Charlatanism, when it is successful, is then, in magic as in everything else, a great instrument of power. To fascinate the mob cleverly, is not that already to dominate it? The poor devils of sorcerers who in the Middle Ages stupidly got themselves burnt alive had not, it is easy to see, a great empire on others. Joan of Arc was a magician at the head of her armies, and at Rouen the poor girl was not even a witch. She only knew how to pray, and how to fight, and the prestige which surrounded her ceased as soon as she was in chains. Does history tell us that the King of France demanded her release? That the French nobility, the people, the army protested against her condemnation? The Pope, whose eldest son was the King of France, did he excommunicate the executioners of the Maid of Orleans? No, nothing of all that! Joan of Arc was a sorceress for every one as soon as she ceased to be a magician, and it was certainly not the English alone who burned her. When one exercises an apparently superhuman power, one must exercise it always, or resign oneself to perish. The world always avenges itself in a cowardly way for having believed too much, admired too much, and above all, obeyed too much.

We only understand magic power in its application to great matters. If a true practical magician does not make himself master of the world, it is that he disdains it. To what, then, would he degrade his sovereign power? 'I will give thee all the kingdoms of the world, if thou wilt fall at my feet and worship me,' the Satan of the parable said to Jesus. 'Get thee behind me, Satan,' replied the Saviour; 'for it is written, Thou shalt adore God alone.' . . . 'ELI, ELI, LAMA SABACHTHANI!' was what this sublime and divine adorer of God cried later. If he had replied to Satan: 'I will not adore thee, and it is thou who wilt fall at my feet, for I bid thee in the name of intelligence and eternal reasons,' he would not have consigned his holy and noble life to the most frightful of all tortures. The Satan of the mountain was indeed cruelly avenged!

The ancients called practical magic the sacerdotal and royal art, and one remembers that the magi were the masters of primitive civilization, because they were the masters of all the science of their time.

To know is to be able when one dares to will.

The first science of the practical qabalist, or the magus, is the knowledge of men. Phrenology, psychology, chiromancy, the observation of tastes and of movement, of the sound of the voice and of either sympathetic or antipathetic impressions, are branches of this art, and the ancients were not ignorant of them. Gall and Spurzheim in our days have rediscovered phrenology. Lavater, following Porta, Cardan,

Taisnier, Jean Belot and some others have divined anew rather than rediscovered the science of psychology; cheiromancy is still occult, and one scarcely finds traces of it in the quite recent and very interesting work of d'Arpentigny. In order to have sufficient notions of it, one must remount to the qabalistic sources themselves from which the learned Cornelius Agrippa drew water. It is, then, convenient to say a few words on the subject while waiting for the work of our friend Desbarrolles.

The hand is the instrument of action in man: it is, like the face, a sort of synthesis of the nervous system, and should also have features and physiognomy. The character of the individual is traced there by undeniable signs. Thus, among hands, some are laborious, some are idle, some square and heavy, others insinuating and light. Hard and dry hands are made for strife and toil, soft and damp hands ask only for pleasure. Pointed fingers are inquisitive and mystical, square fingers mathematical, spatulated fingers obstinate and ambitious.

The thumb, pollex, the finger of force and power, corresponds in the qabalistic symbolism to the first letter of the name of Jehovah. This finger is then a synthesis of the hand: if it is strong, the man is morally strong; if it is weak, the man is weak. It has three phalanges, of which the first is hidden in the palm of the hand, as the imaginary axis of the world traverses the thickness of the earth. This first phalanx corresponds to the physical life, the second to the intelligence, the third to the will. Greasy and thick palms denote sensual tastes and great force of physical life; a thumb which is long, especially in its last phalanx, reveals a strong will, which may go as far as despotism; short thumbs, on the contrary, show characters gentle and easily controlled.

The habitual folds of the hand determine its lines. These lines are, then, the traces of habits, and the patient observer will know how to recognize them and how to judge them. The man whose hand folds badly is clumsy or unhappy. The hand has three principal functions: to grasp, to hold, and to handle. The subtlest hands seize and handle best; hard and strong hands hold longer. Even the lightest wrinkles bear witness to the habitual sensations of the organ. Each finger has, besides, a special function from which it takes its name. We have already spoken of the thumb; the index is the finger which points out, it is that of the word and of prophecy; the medius dominates the whole hand, it is that of destiny; the ring-finger is that of alliances and of honours: cheiromancers have consecrated it to the sun; the little finger is insinuating and talkative, at least, so say simple folk and nursemaids, whose little finger tells them so much. The hand has seven protuberances which the qabalists, following natural analogies, have attributed to the seven planets: that of the thumb, to Venus; that of the index, to Jupiter; that of the medius, to Saturn; that of the ring-finger, to the Sun; that of the little finger, to Mercury; the two others to Mars and to the Moon. According to their form and their predominance, they judged the inclinations, the aptitudes, and consequently the probable destinies, of the individuals who submitted themselves to their judgment.

There is no vice which does not leave its trace, no virtue which has not its sign. Thus, for the trained eyes of the observer, no hypocrisy is possible. One will understand that such a science is already a power indeed sacerdotal and royal.

The prediction of the principal events of life is already possible by means of the numerous analogical probabilities of this observation: but there exists a faculty called that of presentiments or sensitivism. Events exist often in their causes before realizing themselves in action; sensitives see in advance the effects in the causes. Previous to all great events, there have been most astonishing predictions. In the reign of Louis Philippe we heard sleep-wakers and ecstatics announce the return of the Empire, and specify the date of its coming. The Republic of 1848 was clearly announced in the prophecy of Orval, which dated at least from 1830 and which we strongly suspect to be, like those works attributed to the brothers Olivarius, the posthumous work of Mlle Lenormand. This is a matter of little importance in this thesis.

That magnetic light which causes the future to appear, also causes things at present existing, but hidden, to be guessed; as it is the universal life, it is also the agent of human sensibility, transmitting to some the sickness or the health of others, according to the fatal influence of contracts, or the laws of the will. It is that which explains the power of benedictions and of bewitchments so clearly recognized by the great adepts, and above all by the wonderful Paracelsus. An acute and judicious critic, Mr. Ch. Fauvety, in an article published by the *Revue philosophique et religieuse,* appreciates in a remarkable manner the advanced works of Paracelsus, of Pomponacius, of Goglienus, of Crollius, and of Robert Fludd on magnetism. But what our learned friend and collaborator studies only as a philosophical curiosity, Paracelsus and his followers practised without being very anxious that the world should understand it; for it was for them one of those traditional secrets with regard to which silence is necessary, and which it is sufficient to indicate to those who know, leaving always a veil upon the truth for the ignorant.

Now here is what Paracelsus reserved for initiates alone, and what we have understood through deciphering the qabalistic characters, and the allegories of which he makes use in his work:

The human soul is material; the divine *mens* is offered to it to immortalize it and to make it live spiritually and individually, but its natural substance is fluidic and collective.

There are, then, in man, two lives: the individual or reasonable life, and the common or instinctive life. It is by this latter that one can live in the bodies of others, since the universal soul, of which each nervous organism has a separate consciousness, is the same for all.

We live in a common and universal life in the embryonic state, in ecstasy, and in sleep. In sleep, in fact, reason does not act, and logic, when it mingles in our dreams, only does so by chance, in accordance with the accidents of purely physical reminiscences.

In dreams, we have the consciousness of the universal life; we mingle ourselves with water, fire, air, and earth; we fly like birds; we climb like squirrels; we crawl like serpents; we are intoxicated with astral light; we plunge into the common reservoir, as happens in a more complete manner in death; but then (and it is thus that Paracelsus explains the mysteries of the other life) the wicked, that is to say, those who have allowed themselves to be dominated by the instincts of the brute to the prejudice of human reason, are drowned in the ocean of the common life with all the anguish of eternal death; the others swim upon it, and enjoy for ever the riches of that fluid gold which they have succeeded in dominating.

This identity of all physical life permits the stronger souls to possess themselves of the existence of the others, and to make auxiliaries of them; it explains sympathetic currents either near or distant, and gives the whole secret of occult medicine, because the principle of this medicine is the grand hypothesis of universal analogies, and, attributing all the phenomena of physical life to the universal agent, teaches that one must act upon the astral body in order to react upon the material visible body; it teaches also that the essence of the astral light is a double movement of attraction and repulsion; just as human bodies attract and repel one another, they can also absorb themselves, extend one into another, and make exchanges; the ideas or imaginations of one can influence the form of the other, and subsequently react upon the exterior body.

Thus are produced the so strange phenomena of maternal impressions, thus the neighbourhood of invalids gives bad dreams, and thus the soul breathes in something unwholesome when in the company of fools and knaves.

One may remark that in boarding-schools the children tend to assimilate in physiognomy; each place of education has, so to speak, a family air which is peculiar to it. In orphan schools conducted by nuns all the girls resemble each other, and all take on that obedient and effaced physiognomy which characterizes ascetic education. Men become handsome in the school of enthusiasm, of the arts, and of glory; they become ugly in prison, and of sad countenance in seminaries and in convents.

Here it will be understood we leave Paracelsus, in order that we may investigate the consequences and applications of his ideas, which are simply those of the ancient magi, and to study the elements of that physical Qabalah which we call magic.

According to the qabalistic principles formulated by the school of Paracelsus, death is nothing but a slumber, ever growing deeper and more definite, a slumber which it would not be impossible to stop in its early stages by exercising a powerful action of will on the astral body as it breaks loose, and by recalling it to life through some powerful interest or some dominating affection. Jesus expressed the same thought when He said to the daughter of Jairus: 'The maiden is not dead, but sleepeth'; and of Lazarus: 'Our friend is fallen asleep, and I go to wake him.' To express this resurrectionist system in such a manner as not to offend common sense, by which we mean generally-held opinions, let us say that death, when there is no destruction or essential alteration of the physical organs, is always preceded by a lethargy of varying duration. (The resurrection of Lazarus, if we could admit it as a scientific fact, would prove that this state may last for four days.)[1]

Let us now come to the secret of the Great Work, which we have given only in Hebrew, without vowel points, in *Transcendental Magic*. Here is the complete text in Latin, as one finds it on page 144

---

[1] It will be objected that Lazarus stank, but this is a thing which happens frequently to healthy people, as well as to sick men, who recover in spite of it. Besides, in the Gospel story, it is one of the bystanders who says that Lazarus 'by this time stinketh, for he hath been dead four days'. One may then attribute this remark to imagination.—E.L. Rather to the arrogance of the *à priori* reasoner.—A.C.

of the *Sepher Yetzirah*, commented by the alchemist Abraham (Amsterdam, 1642):

Semita XXXI

Vocatur intelligentia perpetua; et quare vocatur ita? Eo quod ducit motum solis et lunae juxta constitutionem eorum; utrumque in orbe sibi conveniente.

Rabbi Abraham F∴ D∴ dicit:

Semita trigesima prima vocatur intelligentia perpetua: et illa ducit solem et lunam et reliquas stellas et figuras, unum quodque in orbe suo, et impertit omnibus creatis juxta dispositionem ad signa et figuras.

Here is the English translation of the Hebrew text which we have transcribed in our ritual:

'The thirty-first path is called the perpetual intelligence; and it governs the sun and the moon, and the other stars and figures, each in its respective orb. And it distributes what is needful to all created things, according to their disposition to the signs and figures.'

This text, one sees, is still perfectly obscure for whoever is not acquainted with the characteristic value of each of the thirty-two paths. The thirty-two paths are the ten numbers and the twenty-two hieroglyphic letters of the Qabalah. The thirty-first refers to ש, which represents the magic lamp, or the light between the horns of Baphomet. It is the qabalistic sign of the OD, or astral light, with its two poles, and its balanced centre. One knows that in the language of the alchemist the sun signifies gold, the moon silver, and that the other stars or planets refer to the other metals. One should now be able to understand the thought of the Jew Abraham.

The secret fire of the masters of alchemy was, then, electricity; and there is the better half of their grand arcanum; but they knew how to equilibrate its force by a magnetic influence which they concentrated in their athanor. This is what results from the obscure dogmas of Basil Valentine, of Bernard Trevisan, and of Henry Khunrath, who, all of them, pretend to have worked the transmutation, like Raymond Lully, like Arnaud de Villeneuve, and like Nicholas Flamel.

The universal light, when it magnetizes the worlds, is called astral light; when it forms the metals, one calls it azoth, or philosophical mercury; when it gives life to animals, it should be called animal magnetism.

The brute is subject to the fatalities of this light; man is able to direct it.

It is the intelligence which, by adapting the sign to the thought, creates forms and images.

The universal light is like the divine imagination, and this world, which changes ceaselessly, yet ever remaining the same with regard to the laws of its configuration, is the vast dream of God.

Man formulates the light by his imagination; he attracts to himself the light in sufficient quantities to give suitable forms to his thoughts and even to his dreams; if this light overcomes him, if he drowns his understanding in the forms which he evokes, he is mad. But the fluidic atmosphere of madmen is often a poison for tottering reason and for exalted imaginations.

The forms which the over-excited imagination produces in order to lead astray the understanding, are as real as photographic images. One could not see what does not exist. The phantoms of dreams, and even the dreams of the waking, are then real images which exist in the light.

There exist, besides these, contagious hallucinations. But we here affirm something more than ordinary hallucinations.

If the images attracted by diseased brains are in some sense real, can they not throw them without themselves, as real as they receive them?

These images projected by the complete nervous organism of the medium, can they not affect the complete organism of those who, voluntarily or not, are in nervous sympathy with the medium?

The things accomplished by Mr. Home prove that all this is possible.

Now, let us reply to those who think that they see in these phenomena manifestations of the other world and facts of necromancy.

We shall borrow our answer from the sacred book of the qabalists, and in this our doctrine is that of the rabbis who compiled the Zohar.

Axiom

The spirit clothes itself to descend, and strips itself to rise.
In fact:
Why are created spirits clothed with bodies?

It is that they must be limited in order to have a possible existence. Stripped of all body, and become consequently without limit, created spirits would lose themselves in the infinite, and from lack of the power to concentrate themselves somewhere, they would be dead and impotent everywhere, lost as they would be in the immensity of God.

All created spirits have, then, bodies, some subtler, some grosser, according to the surroundings in which they are called to live.

The soul of a dead man would, then, not be able to live in the atmosphere of the living, any more than we can live in earth or in water.

For an airy, or rather an ethereal, spirit, it would be necessary to have an artificial body similar to the apparatus of our divers, in order that it might come to us.

All that we can see of the dead are the reflections which they have left in the atmospheric light, light whose imprints we evoke by the sympathy of our memories.

The souls of the dead are above our atmosphere. Our respirable air becomes earth for them. This is what the Saviour declares in His Gospel, when He makes the soul of a saint say:

'Now the great abyss is established between us, and those who are above can no longer descend to those who are below.'

The hands which Mr. Home causes to appear are, then, composed of air coloured by the reflection which his sick imagination attracts and projects.[1]

One touches them as one sees them; half illusion, half magnetic and nervous force.

These, it seems to us, are very precise and very clear explanations.

Let us reason a little with those who support the theory of apparitions from another world:

Either those hands are real bodies, or they are illusions.

If they are bodies, they are, then, not spirits.

If they are illusions produced by mirages, either in us, or outside ourselves, you admit my argument.

Now, one remark!

It is that all those who suffer from luminous congestion or contagious somnambulism, perish by a violent or, at least, a sudden death.

It is for this reason that one used to attribute to the devil the power of strangling sorcerers.

---

[1] 'The luminous agent being also that of heat, one understands the sudden variations of temperature occasioned by the abnormal projections or sudden absorptions of the light. There follows a sudden atmospheric perturbation, which produces the noises of storms, and the creaking of woodwork.'—E.L.

The excellent and worthy Lavater habitually evoked the alleged spirit of Gablidone.

He was assassinated.

A lemonade-seller of Leipzig, Schroepfer, evoked the animated images of the dead. He blew out his brains with a pistol.

One knows what was the unhappy end of Cagliostro.

A misfortune greater than death itself is the only thing that can save the life of these imprudent experimenters.

They may become idiots or madmen, and then they do not die, if one watches over them with care to prevent them from committing suicide.

Magnetic maladies are the road to madness; they are always born from the hypertrophy or atrophy of the nervous system.

They resemble hysteria, which is one of their varieties, and are often produced, either by excesses of celibacy, or those of exactly the opposite kind.

One knows how closely connected with the brain are the organs charged by Nature with the accomplishment of her noblest work: those whose object is the reproduction of being.

One does not violate with impunity the sanctuary of Nature.

Without risking his own life, no one lifts the veil of the great Isis.

Nature is chaste, and it is to chastity that she gives the key of life.

To give oneself up to impure loves is to plight one's troth to death.

Liberty, which is the life of the soul, is only preserved in the order of Nature. Every voluntary disorder wounds it, prolonged excess murders it.

Then, instead of being guided and preserved by reason, one is abandoned to the fatalities of the ebb and flow of magnetic light.

The magnetic light devours ceaselessly, because it is always creating, and because, in order to produce continually, one must absorb eternally.

Thence come homicidal manias and temptations to commit suicide.

Thence comes that spirit of perversity which Edgar Poe has described in so impressive and accurate a manner, and which Mr. de Mirville would be right to call the devil.

The devil is the giddiness of the intelligence stupefied by the irresolution of the heart.

It is a monomania of nothingness, the lure of the abyss; independently of what it may be according to the decisions of the Catholic, Apostolic, and Roman faith, which we have not the temerity to touch.

As to the reproduction of signs and characters by that universal fluid, which we call astral light, to deny its possibility would be to take little account of the most ordinary phenomena of Nature.

The mirage in the steppes of Russia, the palace of Morgan le Fay, the figures printed naturally in the heart of stones which Gaffarel calls *gamahés,* the monstrous deformities of certain children caused by impressions of the nightmares of their mothers, all these phenomena and many others prove that the light is full of reflections and images which it projects and reproduces according to the evocations of the imagination, of memory, or of desire. Hallucination is not always an objectless reverie: as soon as every one sees a thing it is certainly visible; but if this thing is absurd one must rigorously conclude that everybody is deceived or hallucinated by a real appearance.

To say (for example) that in the magnetic parties of Mr. Home real and living hands come out of the tables, true hands which some see, others touch, and by which still others feel themselves touched without seeing them, to say that these really corporeal hands are hands of spirits, is to speak like children or madmen; it implies a contradiction in terms. But to deem that such or such apparitions, such or such sensations, are produced, is simply to be sincere, and to mock the mockery of the normal man, even when these normal men are as witty as this or that editor of this or that comic journal.

These phenomena of the light which produce apparitions always appear at epochs when humanity is in labour. They are phantoms of the delirium of the world-fever; it is the hysteria of a bored society. Virgil tells us in fine verse that in the time of Caesar Rome was full of spectres; in the time of Vespasion the gates of the Temple of Jerusalem opened of themselves, and a voice was heard crying, 'The gods depart'. Now, when the gods depart, the devils return. Religious feeling transforms itself into superstition when faith is lost; for souls need to believe, because they thirst for hope. How can faith be lost? How can science doubt the infinite harmony? Because the sanctuary of the absolute is always closed for the majority. But the kingdom of truth, which is that of God, suffers violence, and the violent must take it by force. There exists a dogma, there exists a key, there exists a sublime tradition; and this dogma, this key, this tradition is transcendental magic. There only are found the absolute of knowledge and the eternal bases of law, guardian against all madness, all superstition and all error, the Eden of the intelligence, the ease of the heart, and the peace of the soul. We do not say this in the hope of convincing the scoffer, but only to guide the seeker. Courage and good hope to him; he will surely find, since we ourselves have found.

The magical dogma is not that of the mediums. The mediums who dogmatize can teach nothing but anarchy, since their inspiration is drawn from a disordered exaltation. They are always predicting disasters; they deny hierarchical authority; they pose, like Vintras, as sovereign pontiffs. The initiate, on the contrary, respects the hierarchy before all, he loves and preserves order, he bows before sincere beliefs, he loves all signs of immortality in faith, and of redemption by charity, which is all discipline and obedience. We have just read a book published under the influence of astral and magnetic intoxication, and we have been struck by the anarchical tendencies with which it is filled under a great appearance of benevolence and religion. At the head of this book one sees the symbol, or, as the magi call it, *the signature*, of the doctrines which it teaches. Instead of the Christian Cross, symbol of harmony, alliance and regularity, one sees the tortuous tendrils of the vine, jutting from its twisted stem, images of hallucination and of intoxication.

The first ideas set forth by this book are the climax of the absurd. The souls of the dead, it says, are everywhere, and nothing any longer hems them in. It is an infinite overcrowded with gods, returning the one into the other. The souls can and do communicate with us by means of tables and hats. And so, no more regulated instruction, no more priesthood, no more Church, delirium set upon the throne of truth, oracles which write for the salvation of the human race the word attributed to Cambronne, great men who leave the serenity of their eternal destinies to make our furniture dance, and to hold with us conversations like those which Beroalde de Verville[1] makes them hold, in *Le Moyen de Parvenir*. All this is a great pity; and yet, in America, all this is spreading like an intellectual plague. Young America raves, she has fever; she is, perhaps, cutting her teeth. But France! France to accept such things! No, it is not possible, and it is not so. But while they refuse the doctrines, serious men should observe the phenomena, remain calm in the midst of the agitations of all the fanaticisms (for incredulity also has its own), and judge after having examined.

To preserve one's reason in the midst of madmen, one's faith in the

---

[1] Born in 1538—died in 1612. Author of *Le Moyen de Parvenir*. The Bibliophile Jacob suggests that Verville stole his *Moyen de Parvenir* from a lost book of Rabelais. Verville was a Canon of St. Gatien, Tours, and is associated with Tours and Touraine. Balzac's *Contes Drôlatiques* are deemed to have been more inspired by Verville than by Rabelais. —A.C.

midst of superstitions, one's dignity in the midst of buffoons, and one's independence among the sheep of Panurge, is of all miracles the rarest, the finest, and the most difficult to accomplish.

CHAPTER IV

Fluidic Phantoms and their Mysteries

THE ancients gave different names to these: larvae, lemures (empuses). They loved the vapour of shed blood, and fled from the blade of the sword.

Theurgy evoked them, and the Qabalah recognized them under the name of elementary spirits.

They were not spirits, however, for they were mortal.

They were fluidic coagulations which one could destroy by dividing them.

They were a sort of animated mirages, imperfect emanations of human life. The traditions of Black Magic say that they were born owing to the celibacy of Adam. Paracelsus says that the vapours of the blood of hysterical women people the air with phantoms; and these ideas are so ancient, that we find traces of them in Hesiod, who expressly forbids that linen, stained by a pollution of any sort, should be dried before a fire.

Persons who are obsessed by phantoms are usually exalted by too rigorous celibacy, or weakened by excesses.

Fluidic phantoms are the abortions of the vital light; they are plastic media without body and without spirit, born from the excesses of the spirit and the disorders of the body.

These wandering media may be attracted by certain degenerates who are fatally sympathetic to them, and who lend them at their own cost a factitious existence of a more or less durable kind. They then serve as supplementary instruments to the instinctive volitions of these degenerates: never to cure them, always to send them farther astray, and to hallucinate them more and more.

If corporeal embryos can take the forms which the imagination of their mothers gives them, the wandering fluidic embryos ought to be prodigiously variable, and to transform themselves with an astonishing facility. Their tendency to give themselves a body in order to attract a soul, makes them condense and assimilate naturally the corporeal molecules which float in the atmosphere.

Thus, by coagulating the vapour of blood, they remake blood, that blood which hallucinated maniacs see floating upon pictures or statues. But they are not the only ones to see it. Vintras and Rose Tamisier are neither impostors nor myopics; the blood really flows; doctors examine it, analyse it; it is blood, real human blood: whence comes it? Can it be formed spontaneously in the atmosphere? Can it naturally flow from a marble, from a painted canvas or a host? No, doubtless; this blood did once circulate in veins, then it has been shed, evaporated, dried, the serum has been turned into vapour, the globules into impalpable dust, the whole has floated and whirled into the atmosphere, and has then been attracted into the current of a specified electromagnetism. The serum has again become liquid; it has taken up and imbibed anew the globules which the astral light has coloured, and the blood flows.

Photography proves to us sufficiently that images are real modifications of light. Now, there exists an accidental and fortuitous photography which makes durable impressions of mirages wandering in the atmosphere, upon leaves of trees, in wood, and even in the heart of stones: thus are formed those natural figures to which Gaffarel has consecrated several pages in his book of *Curiosités inouies*, those stones to which he attributes an occult virtue, and which he calls *gamahés*; thus are traced those writings and drawings which so greatly astonish the observers of fluidic phenomena. They are astral photographs traced by the imagination of the mediums with or without the assistance of the fluidic larvae.

The existence of these larvae has been demonstrated to us in a peremptory manner by a rather curious experience. Several persons, in order to test the magic power of the American Home, asked him to summon up relations which they pretended they had lost, but who, in reality, had never existed. The spectres did not fail to reply to this appeal, and the phenomena which habitually followed the evocations of the medium were fully manifested.

This experience is sufficient of itself to convict of tiresome credulity and of formal error those who believe that spirits intervene to produce these strange phenomena. That the dead may return, it is above all necessary that they should have existed, and demons would not so easily be the dupes of our mystifications.

Like all Catholics, we believe in the existence of spirits of darkness, but we know also that the divine power has given them the dark-ness for an eternal prison, and that the Redeemer saw Satan fall from Heaven like lightning. If the demons tempt us, it is by the voluntary complicity of our passions, and it is not permitted to them to make head against the empire of God, and by stupid and useless manifesta-tions to disturb the eternal order of Nature.

The diabolical signatures and characters, which are produced without the knowledge of the medium, are evidently not proofs of a tacit or formal pact between these degenerates and intelligences of the abyss. These signs have served from the beginning to express astral vertigo, and remain in a state of mirage in the reflection of the divulged light. Nature also has its recollections, and sends to us the same signs to correspond to the same ideas. In all this, there is nothing either super-natural or infernal.

'How? Do you want me to admit,' said to us the Curé Charvoz, the first vicar of Vintras, 'that Satan dares to impress his hideous stigmata upon consecrated materials, which have become the actual body of Jesus Christ?' We declared immediately that it was equally impossible for us to pronounce in favour of such a blasphemy; and yet, as we demonstrated in our articles in the *Estafette*, the signs printed in bleeding characters upon the hosts of Vintras, regularly consecrated by Charvoz, were those which, in Black Magic, are absolutely recog-nized for the signatures of demons.

Astral writings are often ridiculous or obscene. The pretended spirits, when questioned on the greater mysteries of Nature, often reply by that coarse word which became, so they say, heroic on one occasion, in the military mouth of Cambronne. The drawings which pencils will trace if left to their own devices very often reproduce shapeless phalli, such as the anaemic hooligan, as one might pic-turesquely call him, sketches on the hoardings as he whistles, a further proof of our hypothesis, that wit in no way presides at those manifesta-tions, and that it would be above all sovereignly absurd to recognize in them the intervention of spirits released from the bondage of matter.

The Jesuit, Paul Saufidius, who has written on the manners and customs of the Japanese, tells us a very remarkable story. A troop of Japanese pilgrims one day, as they were traversing a desert, saw coming towards them a band of spectres whose number was equal to that of the pilgrims, and which walked at the same pace. These spectres, at first without shape, and like larvae, took on as they approached all the appearance of the human body. Soon they met the pilgrims, and mingled with them, gliding silently between their ranks. Then the Japanese saw themselves double, each phantom having become the perfect image and, as it were, the mirage of each pilgrim. The Japanese were afraid, and prostrated themselves, and the bonze who was conducting them began to pray for them with great contortions and great cries. When the pilgrims rose up again, the phantoms had disappeared, and the troop of devotees was able to continue its path in peace. This phenomenon, whose truth we do not doubt, presents the double characters of a mirage, and of a sudden projection of astral larvae, occasioned by the heat of the atmosphere, and the fanatical exhaustion of the pilgrims.

Dr. Brierre de Boismont, in his curious treatise, *Traité des hallucinations*, tells us that a man, perfectly sane, who had never had visions, was tormented one morning by a terrible nightmare: he saw in his room a mysterious ape horrible to behold, who gnashed his teeth upon him, and gave himself over to the most hideous contortions. He woke with a start, it was already day; he jumped from his bed, and was frozen with terror on seeing, really present, the frightful object of his dream. The monkey was there, the exact image of the monkey of the nightmare, equally absurd, equally terrible, even making the same grimaces. He could not believe his eyes; he remained nearly half an hour motionless, observing this singular phenomenon, and asking himself whether he was delirious or mad. Ultimately, he approached the phantasm to touch it, and it vanished.

Cornelius Gemma, in his *Histoire Critique Universelle*, says that in the year 454, in the island of Candia, the phantom of Moses appeared to some Jews on the sea-side; on his forehead he had luminous horns, in his hand was his blasting rod; and he invited them to follow him, showing them with his finger the horizon in the direction of the Holy Land. The news of this prodigy spread abroad, and the Israelites rushed towards the shore in a mob. All saw, or pretended to see, the marvellous apparition: they were, in number, twenty thousand, according to the chronicler, whom we suspect to be slightly exaggerating in this respect. Immediately heads grow hot, and imaginations wild; they believe in a miracle more startling than was of old the passage of the Red Sea. The Jews form in a close column, and run towards the sea; the rear ranks push the front ranks frantically: they think they see the pretended Moses walk upon the water. A shocking disaster resulted: almost all that multitude was drowned, and the hallucination was only extinguished with the life of the greater number of those unhappy visionaries.

Human thought creates what it imagines; the phantoms of superstition project their deformities on the astral light, and live upon the same terrors which give them birth. That black giant which reaches its wings from east to west to hide the light from the world, that monster who devours souls, that frightful divinity of ignorance and fear—in a word, the devil—is still, for a great multitude of children of all ages, a frightful reality. In our *Transcendental Magic* we represented him as the shadow of God, and in saying that, we still hid the half of our thought: God is light without shadow. The devil is only the shadow of the phantom of God!

The phantom of God! that last idol of the earth; that anthropomorphic spectre which maliciously makes himself invisible; that finite personification of the infinite; that invisible whom one cannot see without dying—without dying at least to intelligence and to reason, since in order to see the invisible, one must be mad; the phantom of Him who has no body; the confused form of Him who is without form and without limit; it is in *that* that, without knowing it, the greater number of believers believe. He who *is* essentially, purely, spiritually, without being either absolute being, or an abstract being, or the collection of beings, the intellectual infinite in a word, is so difficult to imagine! Besides, every imagination makes its creator an idolater; he is obliged to believe in it, and worship it. Our spirit should be silent before Him, and our heart alone has the right to give Him a name: Our Father!

SECOND BOOK

CHAPTER I

Theory of the Will

HUMAN life and its innumerable difficulties have for object, in the ordination of eternal wisdom, the education of the will of man.

The dignity of man consists in doing what he will, and in willing the good, in conformity with the knowledge of truth.

The good in conformity with the true, is the just.

Justice is the practice of reason.

Reason is the word of reality.

Reality is the science of truth.

Truth is idea identical with being.

Man arrives at the absolute idea of being by two roads, experience and hypothesis.

Hypothesis is probable when it is necessitated by the teachings of experience; it is improbable or absurd when it is rejected by this teaching.

Experience is science, and hypothesis is faith.

True science necessarily admits faith; true faith necessarily reckons with science.

Pascal blasphemed against science, when he said that by reason man could not arrive at the knowledge of any truth.

In fact, Pascal died mad.

But Voltaire blasphemed no less against science, when he declared that every hypothesis was absurd, and admitted for the rule of reason only the witness of the senses.

Moreover, the last word of Voltaire was this contradictory formula: 'GOD AND LIBERTY'.

God! that is to say, a Supreme Master, excludes every idea of liberty as the school of Voltaire understood it.

And Liberty, by which is meant an absolute independence of any master, excludes all idea of God.

The word GOD expresses the supreme personification of law, and by consequence, of duty; and if by the word LIBERTY, you are willing to accept our interpretation, THE RIGHT OF DOING ONE'S DUTY, we in our turn will take it for a motto; and we shall repeat, without contradiction and without error: 'GOD AND LIBERTY'.

As there is no liberty for man but in the order which results from the true and the good, one may say that the conquest of liberty is the great work of the human soul. Man, by freeing himself from his evil passions and their slavery, creates himself, as it were, a second time. Nature made him living and suffering; he makes himself happy and immortal; he thus becomes the representative of divinity upon earth, and (relatively) exercises its almighty power.

Axiom I

Nothing resists the will of man, when he knows the truth, and wills the good.

Axiom II

To will evil, is to will death. A perverse will is a beginning of suicide.

Axiom III

To will good with violence, is to will evil, for violence produces disorder, and disorder produces evil.

Axiom IV

One can, and one should, accept evil as the means of good; but one must never will it or do it, otherwise one would destroy with one hand what one builds with the other. Good faith never justifies bad means; it corrects them when one undergoes them, and condemns them when one takes them.

Axiom V

To have the right to possess always, one must will patiently and long.

Axiom VI

To pass one's life in willing what it is impossible to possess always, is to abdicate life and accept the eternity of death.

Axiom VII

The more obstacles the will surmounts, the stronger it is. It is for this reason that Christ glorified poverty and sorrow.

Axiom VIII

When the will is vowed to the absurd, it is reproved by eternal reason.

Axiom IX

The will of the just man is the will of God himself, and the law of Nature.

Axiom X

It is by the will that the intelligence sees. If the will is healthy, the sight is just. God said: 'Let there be light!' and light is; the will says: 'Let the world be as I will to see it!' and the intelligence sees it as the will has willed. This is the meaning of the word, 'So be it,' which confirms acts of faith.

Axiom XI

When one creates phantoms for oneself, one puts vampires into the world, and one must nourish these children of a voluntary nightmare with one's blood, one's life, one's intelligence, and one's reason, without ever satisfying them.

Axiom XII

To affirm and to will what ought to be is to create; to affirm and will what ought not to be, is to destroy.

Axiom XIII

Light[1] is an electric fire put by Nature at the service of the will; it lights those who know how to use it, it burns those who abuse it.

Axiom XIV

The empire of the world is the empire of the light.[1]

Axiom XV

Great intellects whose wills are badly balanced are like comets which are aborted suns.

Axiom XVI

To do nothing is as fatal as to do evil, but it is more cowardly. The most unpardonable of mortal sins is inertia.

Axiom XVII

To suffer is to work. A great sorrow suffered is a progress accomplished. Those who suffer much live more than those who do not suffer.

[1] Meaning again the special 'light' spoken of previously.—A.C.

Axiom XVIII

Voluntary death from devotion is not suicide; it is the apotheosis of the will.

Axiom XIX

Fear is nothing but idleness of the will, and for that reason public opinion scourges cowards.

Axiom XX

Succeed in not fearing the lion, and the lion will fear you. Say to sorrow: 'I will that you be a pleasure, more even than a pleasure, a happiness.'

Axiom XXI

A chain of iron is easier to break than a chain of flowers.

Axiom XXII

Before saying that a man is happy or unhappy, find out what the direction of his will has made of him: Tiberius died every day at Capri, while Jesus proved his immortality and even his divinity on Calvary and upon the Cross.

CHAPTER II

The Power of the Word

IT IS the word which creates forms; and forms in their turn react upon the word, in order to modify it and complete it.

Every word of truth is a beginning of an act of justice.

One asks if man may sometimes be necessarily driven to evil. Yes, when his judgment is false, and consequently his word unjust.

But one is responsible for a false judgment as for a bad action.

What falsifies the judgment is selfishness and its unjust vanities.

The unjust word, unable to realize itself by creation, realizes itself by destruction. It must either slay or be slain.

If it were able to remain without action, it would be the greatest of all disorders, an abiding blasphemy against truth.

Such is that idle word of which Christ has said that one will give account at the Day of Judgment. A jesting word, a comicality which *recreates* and causes laughter, is not an idle word.

The beauty of the word is a splendour of truth. A true word is always beautiful, a beautiful word is always true.

For this reason works of art are always holy when they are beautiful.

What does it matter to me that Anacreon should sing of Bathyllus, if in his verse I hear the notes of that divine harmony which is the eternal hymn of beauty? Poetry is pure as the Sun: it spreads its veil of light over the errors of humanity. Woe to him who would lift the veil in order to perceive things ugly!

The Council of Trent decided that it was permissible for wise and prudent persons to read the books of the ancients, even those which were obscene, on account of the beauty of the form. A statue of Nero or of Heliogabalus made like a masterpiece of Phidias, would it not be an absolutely beautiful and absolutely good work?—and would not he deserve the execration of the whole world who would propose to break it because it was the representation of a monster?

Scandalous statues are those which are badly sculptured, and the Venus of Milo would be desecrated if one placed her beside some of the Virgins which they dare to exhibit in certain churches.

One realizes evil in books of morality ill-written far more than in the poetry of Catullus or the ingenious Allegories of Apuleius.

There are no bad books, except those which are badly conceived and badly executed.

Every word of beauty is a word of truth. It is a light crystallized in speech.

But in order that the most brilliant light may be produced and made visible, a shadow is necessary; and the creative word, that it may become efficacious, needs contradictions. It must submit to the ordeal of negation, of sarcasm, and then to that more cruel yet, of indifference and forgetfulness. The Master said: 'If a corn of wheat fall into the ground and die, it abideth alone; but if it die, it bringeth forth much fruit.'

Affirmation and negation must, then, marry each other, and from their union will be born the practical truth, the real and progressive word. It is necessity which should constrain the workmen to choose for the corner-stone that which they had at first despised and rejected. Let contradiction, then, never discourage men of initiative! Earth is necessary for the ploughshare, and the earth resists because it is in labour. It defends itself like all virgins; it conceives and brings forth slowly like all mothers. You, then, who wish to sow a new plant in the field of intelligence, understand and respect the modesties and reluctances of limited experience and slow-moving reason.

When a new word comes into the world, it needs swaddling clothes and bandages; genius brought it forth, but it is for experience to nourish it. Do not fear that it will die of neglect! Oblivion is for it a favourable time of rest, and contradictions help it to grow. When a sun bursts forth in space it creates worlds or attracts them to itself. A single spark of fixed light promises a universe to space.

All magic is in a word, and that word pronounced qabalistically is stronger than all the powers of Heaven, Earth and Hell. With the name of *Jod Hé Vau Hé*, one commands Nature: kingdoms are conquered in the name of Adonai, and the occult forces which compose the empire of Hermes are one and all obedient to him who knows how to pronounce duly the incommunicable name of Agla.

In order to pronounce duly the great words of the Qabalah, one must pronounce them with a complete intelligence, with a will that nothing checks, an activity that nothing daunts. In magic, to have said is to have done; the word begins with letters, it ends with acts. One does not really will a thing unless one wills it with all one's heart, to the point of breaking for it one's dearest affections; and with all one's forces, to the point of risking one's health, one's fortune, and one's life.

It is by absolute devotion that faith proves itself and constitutes itself. But the man armed with such a faith will be able to move mountains.

The most fatal enemy of our souls is idleness. Inertia intoxicates us and sends us to sleep; but the sleep of inertia is corruption and death. The faculties of the human soul are like the waves of the ocean. To keep them sweet, they need the salt and bitterness of tears: they need the whirlwinds of Heaven: they need to be shaken by the storm.

When, instead of marching upon the path of progress, we wish to have ourselves carried, we are sleeping in the arms of death. It is to us that it is spoken, as to the paralytic man in the Gospel, 'Take up thy bed and walk!' It is for us to carry death away, to plunge it into life.

Consider the magnificent and terrible metaphor of St. John; Hell is a sleeping fire. It is a life without activity and without progress; it is sulphur in stagnation: *stagnum ignis et sulphuris*.

The sleeping life is like the idle word, and it is of that that men will have to give an account in the Day of Judgment.

Intelligence speaks, and matter stirs. It will not rest until it has taken the form given to it by the word. Behold the Christian word, how for these nineteen centuries it has put the world to work! What battles of giants! How many errors set forth and rebutted! How much deceived and irritated Christianity lies at the bottom of Protestantism, from the sixteenth century to the eighteenth! Human egotism, in despair at its defeats, has whipped up all its stupidities in turn. They have re-clothed the Saviour of the world with every rag and with every mocking purple. After Jesus the Inquisitor they have invented the *sans-culotte* Jesus! Measure if you can all the tears and all the blood that have flowed; calculate audaciously all that will yet be shed before the arrival of the Messianic reign of the Man-God who shall submit at once all passions to powers and all powers to justice. THY KINGDOM COME! For nigh on nineteen hundred years, over the whole surface of the earth, this has been the cry of seven hundred million throats, and the Israelites yet await the Messiah! He said that he would come, and come he will. He came to die, and he has promised to return to live.

HEAVEN IS THE HARMONY OF GENEROUS SENTIMENTS
HELL IS THE CONFLICT OF COWARDLY INSTINCTS

When humanity, by dint of bloody and dolorous experience, has truly understood this double truth, it will adjure the Hell of selfishness to enter into the Heaven of devotion and of Christian charity.

The lyre of Orpheus civilized savage Greece, and the lyre of Amphion built Thebes the Mysterious, because harmony is truth. The whole of Nature is harmony. But the Gospel is not a lyre; it is the book of the eternal principles which should and will regulate all the lyres and all the living harmonies of the universe.

While the world does not understand these three words: Truth,

Reason, Justice, and these: Duty, Hierarchy, Society, the revolutionary motto, 'Liberty, Equality, Fraternity,' will be nothing but a threefold lie.

CHAPTER III

Mysterious Influences

NO MIDDLE course is possible. Every man is either good or bad. The indifferent, the lukewarm are not good; they are consequently bad, and the worst of all the bad, for they are imbecile and cowardly. The battle of life is like a civil war; those who remain neutral betray both parties alike, and renounce the right to be numbered among the children of the fatherland.

We all of us breathe in the life of others, and we breathe upon them in some sort a part of our own existence. Good and intelligent men are, unknown to themselves, the doctors of humanity; foolish and wicked men are public poisoners.

There are people in whose company one feels refreshed. Look at that young society woman! She chatters, she laughs, she dresses like everybody else; why, then, is everything in her better and more perfect? Nothing is more natural than her manner, nothing franker and more nobly free than her conversation. Near her everything should be at its ease, except bad sentiments, but near her they are impossible. She does not seek hearts, but draws them to herself and lifts them up. She does not intoxicate, she enchants. Her whole personality preaches a perfection more amiable than virtue itself. She is more gracious than grace, her acts are easy and inimitable, like fine music and poetry. It is of her that a charming woman, too friendly to be her rival, said after a ball: 'I thought I saw the Holy Bible frolicking.'

Now look upon the other side of the sheet! See this other woman who affects the most rigid devotion, and would be scandalized if she heard the angels sing; but her talk is malevolent, her glance haughty and contemptuous; when she speaks of virtue she makes vice lovable. For her God is a jealous husband, and she makes a great merit of not deceiving Him. Her maxims are desolating, her actions due to vanity more than to charity, and one might say after having met her at church: 'I have seen the devil at prayer.'

On leaving the first, one feels one's self full of love for all that is beautiful, good and generous. One is happy to have well said to her all the noble things with which she has inspired you, and to have been approved by her. One says to one's self that life is good, since God has bestowed it on such souls as hers; one is full of courage and of hope. The other leaves you weakened and baffled, or perhaps, what is worse, full of evil designs; she makes you doubt of honour, piety and duty; in her presence one only escapes from weariness by the door of evil desires. One has uttered slander to please her, humiliated one's self to flatter her pride, one remains discontented with her and with one's self.

The lively and certain sentiment of these diverse influences is proper to well-balanced spirits and delicate consciences, and it is precisely that which the old ascetic writers called the power of discerning spirits.

You are cruel consolers, said Job to his pretended friends. It is, in fact, the vicious that afflict rather than console. They have a prodigious tact for finding and choosing the most desperate banalities. Are you weeping for a broken affection? How simple you are! they were playing with you, they did not love you. You admit sorrowfully that your child limps; in friendly fashion they bid you remark that he is a hunchback. If he coughs and that alarms you, they conjure you tenderly to take great care of him, perhaps he is consumptive. Has your wife been ill for a long time? Cheer up, she will die of it!

Hope and work is the message of Heaven to us by the voice of all good souls. Despair and die, Hell cries to us in every word and movement, even in all the friendly acts and caresses of imperfect or degraded beings.

Whatever the reputation of any one may be, and whatever may be the testimonies of friendship that that person may give you, if, on leaving him, you feel yourself less well disposed and weaker, he is pernicious for you: avoid him.

Our double magnetism produces in us two sorts of sympathies. We need to absorb and to radiate turn by turn. Our heart loves contrasts, and there are few women who have loved two men of genius in succession.

One finds peace through the protection which one's own weariness of admiration gives; it is the law of equilibrium; but sometimes even sublime natures are surprised in caprices of vulgarity. Man, said the Abbé Gerbet, is the shadow of a God in the body of a beast: there are in him the friends of the angel and the flatterers of the animal. The angel attracts us; but if we are not on our guard, it is the beast that carries us away: it will even drag us fatally with it when it is a question of beastliness; that is to say, of the satisfactions of that life the nourisher of death, which, in the language of beasts, is called 'real life'. In religion, the Gospel is a sure Guide; it is not so in business, and there are a great many people who, if they had to settle the temporal succession of Jesus Christ, would more willingly come to an agreement with Judas Iscariot than with St. Peter.

One admires probity, said Juvenal, and one leaves it to freeze to death. If such and such a celebrated man, for example, had not scandalously solicited wealth, would one ever have thought of endowing his old muse? Who would have left him legacies?

Virtue has our admiration, our purse owes it nothing, that great lady is rich enough without us. One would rather give to vice, it is so poor!

'I do not like beggars, and I only give to the poor who are ashamed to beg,' said one day a man of wit. 'But what do you give them since you do not know them?' 'I give them my admiration and my esteem, and I have no need to know them to do that.' 'How is it that you need so much money?' they asked another, 'you have no children and no calls on you.' 'I have my poor folk, and I cannot prevent myself from giving them a great deal of money.' 'Make me acquainted with them, perhaps I will give them something too.' 'Oh! you know some of them already, I have no doubt. I have seven who cost me an enormous amount, and an eighth who costs more than the seven others. The seven are the seven deadly sins; the eighth is gambling.'

Another dialogue:

'Give me five francs, sir, I am dying of hunger.' 'Imbecile! you are dying of hunger, and you want me to encourage you in so evil a course? You are dying of hunger, and you have the impudence to admit it. You wish to make me the accomplice of your incapacity, the abetter of your suicide. You want to put a premium on wretchedness. For whom do you take me? Do you think I am a rascal like yourself? . . .

And yet another:

'By the way, old fellow, could you lend me a thousand pounds? I want to seduce an honest woman.' 'Ah! that is bad, but I can never refuse anything to a friend. Here they are. When you have succeeded you might give me her address.' That is what is called in England, and elsewhere, the manners of a gentleman.

'The man of honour who is out of work steals, and does not beg!' replied, one day, Cartouche to a passer-by who asked alms of him. It is as emphatic as the word which tradition associates with Cambronne, and perhaps the famous thief and the great general both really replied in the same manner.

It was that same Cartouche who offered, on another occasion, of his own accord and without it being asked of him, twenty thousand pounds to a bankrupt. One must act properly to one's brothers.

Mutual assistance is a law of nature. To aid those who are like ourselves is to aid ourselves. But above mutual assistance rises a holier and greater law: it is universal assistance, it is charity.

We all admire and love Saint Vincent de Paul, but we have also a secret weakness for the cleverness, the presence of mind, and, above all, the audacity of Cartouche.

The avowed accomplices of our passions may disgust us by humiliating us; at our own risk and peril our pride will teach us how to resist them. But what is more dangerous for us than our hypocritical and hidden accomplices? They follow us like sorrow, await us like the abyss, surround us like infatuation. We excuse them in order to excuse ourselves, defending them in order to defend ourselves, justifying them in order to justify ourselves, and we submit to them finally because we must, because we have not the strength to resist our inclinations, because we lack the will to do so.

They have possessed themselves of our ascendant, as Paracelsus says, and where they wish to lead us we shall go.

They are our bad angels. We know it in the depths of our consciousness; but we put up with them, we have made ourselves their servants that they also may be ours.

Our passions, treated tenderly and flattered, have become slave-mistresses; and those who serve our passions our valets, and our masters.

We breathe out our thoughts and breathe in those of others imprinted in the astral light which has become their electro-magnetic atmosphere: and thus the companionship of the wicked is less fatal to the good than that of vulgar, cowardly, and tepid beings. Strong antipathy warns us easily, and saves us from the contact of gross vices; it is not thus with disguised vices, vices to a certain extent diluted and become almost lovable. An honest woman will experience nothing but disgust in the society of a prostitute, but she has everything to fear from the seductions of a coquette.

One knows that madness is contagious, but the mad are more particularly dangerous when they are amiable and sympathetic. One enters little by little into their circle of ideas, one ends by understanding their exaggerations, while partaking their enthusiasms, one grows accustomed to their logic that has lost its way, one ends by finding that they are not as mad as one thought at first. Thence to believing that they alone are right there is but one step. One likes them, one approves of them, one is as mad as they are.

The affections are free and may be based on reason, but sympathies are of fatalism, and very frequently unreasonable. They depend on the more or less balanced attractions of the magnetic light, and act on men in the same way as upon animals. One will stupidly take pleasure in the society of a person in whom is nothing lovable, because one is mysteriously attracted and dominated by him. And often enough, these strange sympathies began by lively antipathies; the fluids repelled each other at first, and subsequently became balanced.

The equilibrating speciality of the plastic medium of every person is what Paracelsus calls his *ascendant*, and he gives the name of *flagum* to the particular reflection of the habitual ideas of each one in the universal light.

One arrives at the knowledge of the *ascendant* of a person by the sensitive divination of the *flagum*, and by a persistent direction of the will. One turns the active side of one's own ascendant towards the passive side of the ascendant of another when one wishes to take hold of that other and dominate him.

The astral ascendant has been divined by other magi, who gave it the name of *tourbillon* (vortex).

It is, say they, a current of specialized light, representing always the same circle of images, and consequently determined and determining impressions. These vortices exist for men as for stars. 'The stars,' said Paracelsus, 'breathe out their luminous soul, and attract each other's radiation. The soul of the earth, prisoner of the fatal laws of gravitation, frees itself by specializing itself, and passes through the instinct of animals to arrive at the intelligence of man. The active portion of this will is dumb, but it preserves in writing the secrets of Nature. The free part can no longer read this fatal writing without instantaneously losing its liberty. One does not pass from dumb and vegetative contemplation to free vibrating thought without changing one's surroundings and one's organs. Thence comes the forgetfulness which accompanies birth, and the vague reminiscences of our sickly intuitions, always analogous to the visions of our ecstasies and of our dreams.'

This revelation of that great master of occult medicine throws a fierce light on all the phenomena of somnambulism and of divination. There also, for whoever knows how to find it, is the true key of evocation, and of communication with the fluidic soul of the earth.

Those persons whose dangerous influence makes itself felt by a single touch are those who make part of a fluidic association, or who either voluntarily or involuntarily make use of a current of astral light which has gone astray. Those, for example, who live in isolation, deprived of all communication with humanity, and who are daily in fluidic sympathy with animals gathered together in great number, as is ordinarily the case with shepherds, are possessed of the demon whose name is *legion*; in their turn they reign despotically over the fluid souls of the flocks that are confided to their care: consequently their good-will or ill-will makes their cattle prosper or die; and this influence of animal sympathy can be exercised by them upon human plastic mediums which are ill defended, owing either to a weak will or a limited intelligence.

Thus are explained the bewitchments which are habitually made by shepherds, and the still quite recent phenomena of the Presbytery of Cideville.

Cideville is a little village of Normandy, where a few years ago were produced phenomena like those which have since occurred under the influence of Mr. Home. M. de Mirville has studied them carefully, and M. Gougenet Desmousseaux has reprinted all the details in a book, published in 1854, entitled *Moeurs et Pratiques des Démons*. The most remarkable thing in this latter author is that he seems to divine the existence of the plastic medium or the fluidic body. 'We have certainly not two souls,' said he, 'but perhaps we have two bodies.' Everything that he says, in fact, would seem to prove this hypothesis. He saw a shepherd whose fluidic form haunted a Presbytery, and who was wounded at a distance by blows inflicted on his astral larva.

We shall here ask of M. de Mirville and Gougenet Desmousseaux if they take this shepherd for the devil, and if, far or near, the devil such as they conceive him can be scratched or wounded. At that time, in Normandy, the magnetic illnesses of mediums were hardly known, and this unhappy sleep-waker, who ought to have been cared for and cured, was roughly treated and even beaten, not even in his fluidic appearance, but in his proper person, by the Vicar himself. That is, one must agree, a singular kind of exorcism! If those violences really took place, and if they may be imputed to a Churchman whom one considers, and who may be, for all we know, very good and very respectable, let us admit that such writers as M. de Mirville and Gougenet Desmousseaux make themselves not a little his accomplices!

The laws of physical life are inexorable, and in his animal nature man is born a slave to fatality; it is by dint of struggles against his instincts that he may win moral freedom. Two different existences are then possible for us upon the earth; one fatal, the other free. The fatal being is the toy or instrument of a force which he does not direct. Now, when the instruments of fatality meet and collide, the stronger breaks or carries away the weaker; truly emancipated beings fear neither bewitchments nor mysterious influences.

You may reply that an encounter with Cain may be fatal for Abel. Doubtless; but such a fatality is an advantage to the pure and holy victim, it is only a misfortune for the assassin.

Just as among the righteous there is a great community of virtues and merits, there is among the wicked an absolute solidarity of fatal culpability and necessary chastisement. Crime resides in the tendencies of the heart. Circumstances which are almost always independent of the will are the only causes of the gravity of the acts. If fatality had made Nero a slave, he would have become an actor or a gladiator, and would not have burned Rome: would it be to him that one should be grateful for that?

Nero was the accomplice of the whole Roman people, and those who should have prevented them incurred the whole responsibility for the frenzies of this monster. Seneca, Burrhus, Thrasea, Corbulon, theirs is the real guilt of that fearful reign; great men who were either selfish or incapable! The only thing they knew was how to die.

If one of the bears of the Zoological Gardens escaped and devoured several people, would one blame him or his keepers?

Whoever frees himself from the common errors of mankind is obliged to pay a ransom proportional to the sum of these errors: Socrates pays for Aneitus, and Jesus was obliged to suffer a torment whose terror was equal to the whole treason of Judas.

Thus, by paying the debts of fatality, hard-won liberty purchases the empire of the world; it is hers to bind and to unbind. God has put in her hands the keys of Heaven and of Hell.

You men who abandon brutes to themselves wish them to devour you.

The rabble, slaves of fatality, can only enjoy liberty by absolute obedience to the will of free men; they ought to work for those who are responsible for them.

But when the brute governs brutes, when the blind leads the blind, when the leader is as subject to fatality as the masses, what must one expect? What but the most shocking catastrophes? In that we shall never be disappointed.

By admitting the anarchical dogmas of 1789, Louis XVI launched the State upon a fatal slope. From that moment all the crimes of the Revolution weighed upon him alone; he alone had failed in his duty. Robespierre and Marat only did what they had to do. Girondins and Montagnards killed each other in the workings of fatality, and their violent deaths were so many necessary catastrophes; at that epoch there was but one great and legitimate execution, really sacred, really expiatory: that of the King. The principle of royalty would have fallen if that too weak prince had escaped. But a transaction between order and disorder was impossible. One does not inherit from those whom one murders; one robs them; and the Revolution rehabilitated Louis XVI by assassinating him. After so many concessions, so many weaknesses, so many unworthy abasements, that man, consecrated a second time by misfortune, was able at least to say, as he walked to the scaffold: 'The Revolution is condemned, and I am always the King of France!'

To be just is to suffer for all those who are not just, but it is life: to be wicked is to suffer for one's self without winning life: it is to deceive one's self, to do evil, and to win eternal death.

To recapitulate: Fatal influences are those of death. Living influences are those of life. According as we are weaker or stronger in life, we attract or repel witchcraft. This occult power is only too real, but intelligence and virtue will always find the means to avoid its obsessions and its attacks.

CHAPTER IV

Mysteries of Perversity

HUMAN equilibrium is composed of two attractions, one towards death, the other towards life. Fatality is the vertigo which drags us to the abyss; liberty is the reasonable effort which lifts us above the fatal attractions of death. What is mortal sin? It is apostasy from our own liberty; it is to abandon ourselves to the law of inertia. An unjust act is a compact with injustice: now, every injustice is an abdication of intelligence. We fall from that moment under the empire of force whose reactions always crush everything which is unbalanced.

The love of evil and the formal adhesion of the will to injustice are the last efforts of the expiring will. Man, whatever he may do, is more than a brute, and he cannot abandon himself like a brute to fatality. He must choose. He must love. The desperate soul that thinks itself in love with death is still more alive than a soul without love. Activity for evil can and should lead back a man to good, by counter-stroke and by reaction. The true evil, that for which there is no remedy, is inertia.

The abysses of grace correspond to the abysses of perversity. God has often made saints of scoundrels; but He has never done anything with the half-hearted and the cowardly.

Under penalty of reprobation, one must work, one must act. Nature, moreover, sees to this, and if we will not march on with all our courage towards life, she flings us with all her forces towards death. She drags those who will not walk.

A man whom one may call the great prophet of drunkards, Edgar Poe, that sublime madman, that genius of lucid extravagance, has depicted with terrifying reality the nightmares of perversity. . . .

'I killed the old man because he squinted.' 'I did that because I ought not to have done it.'

There is the terrible antistrophe of Tertullian's *Credo quia absurdum.*

To brave God and to insult Him, is a final act of faith. 'The dead praise thee not, O Lord,' said the Psalmist; and we might add if we dared: 'The dead do not blaspheme thee.'

'O my son!' said a father as he leaned over the bed of his child who had fallen into lethargy after a violent access of delirium: 'insult me again, beat me, bite me, I shall feel that you are still alive, but do not rest for ever in the frightful silence of the tomb!'

A great crime always comes to protest against great lukewarmness. A hundred thousand good priests, had their charity been more active, might have prevented the crime of the wretch Verger. The Church has the right to judge, condemn and punish an ecclesiastic who causes scandal; but she has not the right to abandon him to the frenzies of despair and the temptations of misery and hunger.

Nothing is so terrifying as nothingness, and if one could ever formulate the conception of it, if it were possible to admit it, Hell would be a thing to hope for.

This is why Nature itself seeks and imposes expiation as a remedy; that is why chastisement is a chastening, as that great Catholic Count Joseph de Maistre so well understood; this is why the penalty of death is a natural right, and will never disappear from human laws. The stain of murder would be indelible if God did not justify the scaffold; the divine power, abdicated by society and usurped by criminals, would belong to them without dispute. Assassination would then become a virtue when it exercised the reprisals of outraged nature. Private vengeance would protest against the absence of public expiation, and from the splinters of the broken sword of justice anarchy would forge its daggers.

'If God did away with Hell, men would make another in order to defy Him,' said a good priest to us one day. He was right: and it is for that reason that Hell is so anxious to be done away with. Emancipation! is the cry of every vice. Emancipation of murder by the abolition of the pain of death; emancipation of prostitution and infanticide by the abolition of marriage; emancipation of idleness and rapine by the abolition of property. . . . So revolves the whirlwind of perversity until it arrives at this supreme and secret formula: Emancipation of death by the abolition of life!

It is by the victories of toil that one escapes from the fatalities of sorrow. What we call death is but the eternal parturition of Nature. Ceaselessly she re-absorbs and takes again to her breast all that is not born of the spirit. Matter, in itself inert, can only exist by virtue of perpetual motion, and spirit, naturally volatile, can only endure by fixing itself. Emancipation from the laws of fatality by the free adhesion of the spirit to the true and good, is what the Gospel calls the spiritual birth; the re-absorption into the eternal bosom of Nature is the second death.

Unemancipated beings are drawn towards this second death by a fatal gravitation; the one drags the other, as the divine Michel Angelo has made us see so clearly in his great picture of the Last Judgment; they are clinging and tenacious like drowning men, and free spirits must struggle energetically against them, that their flight may not be hindered by them, that they may not be pulled back to Hell.

This war is as ancient as the world; the Greeks figured it under the symbols of Eros and Anteros, and the Hebrews by the antagonism of Cain and Abel. It is the war of the Titans and the Gods. The two armies are everywhere invisible, disciplined and always ready for attack or counter-attack. Simple-minded folk on both sides, astonished at the instant and unanimous resistance that they meet, begin to believe in vast plots cleverly organized, in hidden, all-powerful societies. Eugene Sue invents Rodin;[1] churchmen talk of the Illuminati and of the Freemasons; Wronski dreams of his bands of mystics, and there is nothing true and serious beneath all that but the necessary struggle of order and disorder, of the instincts and of thought; the result of that struggle is balance in progress, and the devil always contributes, despite himself, to the glory of St. Michael.

Physical love is the most perverse of all fatal passions. It is the anarchist of anarchists; it knows neither law, duty, truth nor justice. It would make the maiden walk over the corpses of her parents. It is an irrepressible intoxication; a furious madness. It is the vertigo of fatality seeking new victims; the cannibal drunkenness of Saturn who wishes to become a father in order that he may have more children to devour. To conquer love is to triumph over the whole of Nature. To submit it to justice is to rehabilitate life by devoting it to immortality; thus the greatest works of the Christian revelation are the creation of voluntary virginity and the sanctification of marriage.

While love is nothing but a desire and an enjoyment, it is mortal. In order to make itself eternal it must become a sacrifice, for then it becomes a power and a virtue. It is the struggle of Eros and Anteros which produces the equilibrium of the world.

Everything that over-excites sensibility leads to depravity and crime. Tears call for blood. It is with great emotions as with strong drink; to use them habitually is to abuse them. Now, every abuse of the emotions perverts the moral sense; one seeks them for their own sakes; one sacrifices everything in order to procure them for one's self. A romantic woman will easily become an Old Bailey heroine. She may even arrive at the deplorable and irreparable absurdity of killing herself in order to admire herself, and pity herself, in seeing herself die!

Romantic habits lead women to hysteria and men to melancholia. Manfred, René, Lélia are types of perversity only the more profound in that they argue on behalf of their unhealthy pride, and make poems of their dementia. One asks one's self with terror what monster might be born from the coupling of Manfred and Lélia!

The loss of the moral sense is a true insanity; the man who does not, first of all, obey justice no longer belongs to himself; he walks

[1] Not the sculptor.—A.C.

without a light in the night of his existence; he shakes like one in a dream, a prey to the nightmare of his passions.

The impetuous currents of instinctive life and the feeble resistances of the will form an antagonism so distinct that the qabalists hypothesized the super-foetation of souls: that is to say, they believed in the presence in one body of several souls who dispute it with each other and often seek to destroy it. Very much as the shipwrecked sailors of the *Medusa*, when they were disputing the possession of the too small raft, sought to sink it.

It is certain that, in making one's self the servant of any current whatever, of instincts or even of ideas, one gives up one's personality, and becomes the slave of that multitudinous spirit whom the Gospel calls *legion*. Artists know this well enough. Their frequent evocations of the universal light enervate them. They become *mediums*, that is to say, sick men. The more success magnifies them in public opinion, the more their personality diminishes. They become crotchety, envious, wrathful. They do not admit that any merit, even in a different sphere, can be placed beside theirs; and, having become unjust, they dispense even with politeness. To escape this fatality, really great men isolate themselves from all comradeship, knowing it to be death to liberty. They save themselves by a proud unpopularity from the contamination of the vile multitude. If Balzac had been during his life a man of a clique or of a party, he would not have remained after his death the great and universal genius of our epoch.

The light illuminates neither things insensible nor closed eyes, or at least it only illuminates them for the profit of those who see. The word of Genesis: 'Let there be light!' is the cry of victory with which intelligence triumphs over darkness. This word is sublime in effect because it expresses simply the greatest and most marvellous thing in the world: the creation of intelligence by itself, when, calling its powers together, balancing its faculties, it says: I wish to immortalize myself with the sight of the eternal truth. Let there be light! and there is light. Light, eternal as God, begins every day for all eyes that are open to see it. Truth will be eternally the invention and the creation of genius; it cries: Let there be light! and genius itself is, because light is. Genius is immortal because it understands that light is eternal. Genius contemplates truth as its work because it is the victor of light, and immortality is the triumph of light because it will be the recompense and crown of genius.

But all spirits do not see with justness, because all hearts do not will with justice. There are souls for whom the true light seems to have no right to be. They content themselves with phosphorescent visions, abortions of light, hallucinations of thought; and, loving these phantoms, fear the day which will put them to flight, because they feel that, the day not being made for their eyes, they would fall back into a deeper darkness. It is thus that fools first fear, then calumniate, insult, pursue and condemn the sages. One must pity them, and pardon them, for they know not what they do.

True light rests and satisfies the soul; hallucination, on the contrary, tires it and worries it. The satisfactions of madness are like those gastronomic dreams of hungry men which sharpen their hunger without ever satisfying it. Thence are born irritations and troubles, discouragements and despairs. Life is always a lie to us, say the disciples of Werther, and therefore we wish to die! Poor children, it is not death that you need, it is life. Since you have been in the world you have died every day; is it from the cruel pleasure of annihilation that you would demand a remedy for the annihilation of your pleasures? No, life has never deceived you, you have not yet lived. What you have been taking for life is but the hallucinations and the dreams of the first slumber of death!

All great criminals have hallucinated themselves on purpose; and those who hallucinate themselves on purpose may be fatally led to become great criminals. Our personal light specialized, brought forth, determined by our own over mastering affection, is the germ of our paradise or of our Hell. Each one of us (in a sense) conceives, bears, and nourishes his good or evil angel. The conception of truth gives birth in us to the good genius; intentional untruth hatches and brings up nightmares and phantoms. Every one must nourish his children; and our life consumes itself for the sake of our thoughts. Happy are those who find again immortality in the creations of their soul! Woe unto them who wear themselves out to nourish falsehood and to fatten death! for every one will reap the harvest of his own sowing.

There are some unquiet and tormented creatures whose influence is disturbing and whose conversation is fatal. In their presence one feels one's self irritated, and one leaves their presence angry; yet, by a secret perversity, one looks for them, in order to experience the disturbance and enjoy the malevolent emotions which they give us. Such persons suffer from the contagious maladies of the spirit of perversity.

The spirit of perversity has always for its secret motive the thirst of destruction, and its final aim is suicide. The murderer Éliçabide, on his own confession, not only felt the savage need of killing his relations and friends, but he even wished, had it been possible—he said it in so many words at his trial—*to burst the globe like a cooked chestnut.* Lacenaire, who spent his days in plotting murders, in order to have the means of passing his nights in ignoble orgies or in the excitement of gambling, boasted aloud that he had lived. He called that living, and he sang a hymn to the guillotine, which he called his beautiful betrothed, and the world was full of imbeciles who admired the wretch! Alfred de Musset, before extinguishing himself in drunkenness, wasted one of the finest talents of his century in songs of cold irony and of universal disgust. The unhappy man had been bewitched by the breath of a profoundly perverse woman, who, after having killed him, crouched like a ghoul upon his body and tore his winding sheet. We asked one day, of a young writer of this school, what his literature proved. It proves, he replied frankly and simply, that one must despair and die. What apostleship, and what a doctrine! But these are the necessary and regular conclusions of the spirit of perversity; to aspire ceaselessly to suicide, to calumniate life and nature, to invoke death every day without being able to die. This is eternal Hell, it is the punishment of Satan, that mythological incarnation of the spirit of perversity; the true translation into French of the Greek word *Diabolos,* or devil, is *le pervers*—the perverse.

Here is a mystery which debauchees do not suspect. It is this: one cannot enjoy even the material pleasures of life but by virtue of the moral sense. Pleasure is the music of the interior harmonies; the senses are only its instruments, instruments which sound false in contact with a degraded soul. The wicked can feel nothing, because they can love nothing: in order to love one must be good. Consequently for them everything is empty, and it seems to them that Nature is impotent, because they are so themselves; they doubt everything because they know nothing; they blaspheme everything because they taste nothing; they caress in order to degrade; they drink in order to get drunk; they sleep in order to forget; they wake in order to endure mortal boredom; thus will live, or rather thus will die, every day he who frees himself from every law and every duty in order to make himself the slave of his passions. The world, and eternity itself, become useless to him who makes himself useless to the world and to eternity.

Our will, by acting directly upon our plastic medium, that is to say, upon the portion of astral life which is specialized in us, and which serves us for the assimilation and configuration of the elements necessary to our existence; our will, just or unjust, harmonious or perverse, shapes the medium in its own image and gives it beauty in conformity with what attracts us. Thus moral monstrosity produces physical ugliness; for the astral medium, that interior architect of our bodily edifice, modifies it ceaselessly according to our real or factitious needs. It enlarges the belly and the jaws of the greedy, thins the lips of the miser, makes the glances of impure women shameless, and those of the envious and malicious venomous. When selfishness has prevailed in the soul, the look becomes cold, the features hard: the harmony of form disappears, and according to the absorption or radiant speciality of this selfishness, the limbs dry up or become encumbered with fat. Nature, in making of our body the portrait of our soul, guarantees its resemblance for ever, and tirelessly retouches it. You pretty women who are not good, be sure that you will not long remain beautiful. Beauty is the loan which Nature makes to virtue. If virtue is not ready when it falls due, the lender will pitilessly take back Her capital.

Perversity, by modifying the organism whose equilibrium it destroys, creates at the same time a fatality of needs which urges it to its own destruction, to its death. The less the perverse man enjoys, the more thirsty of enjoyment he is. Wine is like water for the drunkard, gold melts in the hands of the gambler; Messalina tires herself out without being satiated. The pleasure which escapes then changes itself for them into a long irritation and desire. The more murderous are their excesses, the more it seems to them that supreme happiness is at hand. . . . One more bumper of strong drink, one more spasm, one more violence done to Nature. . . . Ah! at last, here is pleasure, here is life . . . and their desire, in the paroxysm of its insatiable hunger, extinguishes itself for ever in death.

PART IV

The Great Practical Secrets or the Realization of Science

INTRODUCTION

THE lofty sciences of the Qabalah and of Magic promise man an exceptional, real, effective, efficient power, and one should regard them as false and vain if they do not give it.

Judge the teachers by their works, said the supreme Master. This rule of judgment is infallible.

If you wish me to believe in what you know, show me what you do.

God, in order to exalt man to moral emancipation, hides Himself from him and abandons to him, after a fashion, the government of the world. He leaves Himself to be guessed by the grandeurs and harmonies of Nature, so that man may progressively make himself perfect by ever exalting the idea that he makes for himself of its author.

Man knows God only by the names which he gives to that Being of beings, and does not distinguish Him but by the images of Him which he endeavours to trace. He is then in a manner the creator of Him who has created him. He believes himself the mirror of God, and by indefinitely enlarging his own mirage, he thinks that he may be able to sketch in infinite space the shadow of Him who is without body, without shadow, and without space.

TO CREATE GOD, TO CREATE ONE'S SELF, TO MAKE ONE'S SELF INDEPENDENT, IMMORTAL AND WITHOUT SUFFERING: there certainly is a programme more daring than the dream of Prometheus. Its expression is bold to the point of impiety, its thought ambitious to the point of madness. Well, this programme is only paradoxical in its form, which lends itself to a false and sacrilegious interpretation. In one sense it is perfectly reasonable and the science of the adepts promises to realize it, and to accomplish it in perfection.

Man, in effect, creates for himself a God corresponding to his own intelligence and his own goodness; he cannot raise his ideal higher than his moral development permits him to do. The God whom he adores is always an enlargement of his own reflection. To conceive the absolute of goodness and justice is to be one's self exceeding just and good.

The moral qualities of the spirit are riches, and the greatest of all riches. One must acquire them by strife and toil. One may bring this objection, the inequality of aptitudes; some children are born with organisms nearer to perfection. But we ought to believe that such organisms result from a more advanced work of Nature, and the children who are endowed with them have acquired them, if not by their own efforts, at least by the consolidated works of the human beings to whom their existence is bound. It is a secret of Nature, and Nature does nothing by chance; the possession of more developed intellectual faculties, like that of money and land, constitutes an indefeasible right of transmission and inheritance.

Yes, man is called to complete the work of his creator, and every instant employed by him to improve himself or to destroy himself, is decisive for all eternity. It is by the conquest of an intelligence eternally clear and of a will eternally just, that he constitutes himself as living for eternal life, since nothing survives injustice and error but the penalty of their disorder. To understand good is to will it, and on the plane of justice to will is to do. For this reason the Gospel tells us that men will be judged according to their works.

Our works make us so much what we are, that our body itself, as we have said, receives the modification, and sometimes the complete change, of its form from our habits.

A form conquered, or submitted to, becomes a providence, or a fatality, for all one's existence. Those strange figures which the Egyptians gave to the human symbols of divinity represent the fatal forms. Typhon has a crocodile's head. He is condemned to eat ceaselessly in order to fill his hippopotamus belly. Thus he is devoted, by his greed and his ugliness, to eternal destruction.

Man can kill or vivify his faculties by negligence or by abuse. He can create for himself new faculties by the good use of those which he has received from Nature. People often say that the affections will not be commanded, that faith is not possible for all, that one does not re-make one's own character. All these assertions are true only for the idle or the perverse. One can make one's self faithful, pious, loving, devoted, when one wishes sincerely to be so. One can give to one's spirit the calm of justness, as to one's will the almighty power of justice. One can reign in Heaven by virtue of faith, on earth by virtue of science. The man who knows how to command himself is king of all Nature.

We are going to state forthwith, in this last book, by what means the true initiates have made themselves the masters of life, how they have overcome sorrow and death; how they work upon themselves and others the transformation of Proteus; how they exercise the divining power of Apollonius; how they make the gold of Raymond Lully and of Flamel; how in order to renew their youth they possess the secrets of Postel the Re-arisen, and those alleged to have been in the keeping of Cagliostro. In short, we are going to speak the last word of magic.

CHAPTER I

Of Transformation—The Wand of Circe—The Bath of Medea—Magic Overcome by its Own Weapons—The Great Arcanum of the Jesuits and The Secret of Their Power.

THE Bible tells us that King Nebuchadnezzar, at the highest point of his power and his pride, was suddenly changed into a beast.

He fled into savage places, began to eat grass, let his beard and hair grow, as well as his nails, and remained in this state for seven years.

In our *Transcendental Magic*, we have said what we think of the mysteries of lycanthropy, or the metamorphosis of men into werewolves.

Everyone knows the fable of Circe and understands its allegory.

The fatal ascendant of one person on another is the true wand of Circe.

One knows that almost all human physiognomies bear a resemblance to one animal or another, that is to say, the *signature* of a specialized instinct.

Now, instincts are balanced by contrary instincts, and dominated by instincts stronger than those.

In order to dominate sheep, the dog plays upon their fear of wolves.

If you are a dog, and you want a pretty little cat to love you, you have only one means to take: to metamorphose yourself into a cat.

But how? By observation, imitation, and imagination. We think that our figurative language will be understood for once, and we recommend this revelation to all who wish to magnetize: it is the deepest of all the secrets of their art.

Here is the formula in technical terms:

'To polarize one's own animal light, in equilibrated antagonism with the contrary pole.'

Or:

'To concentrate in one's self the special qualities of absorption in order to direct their rays towards an absorbing focus, and vice versa.'

This government of our magnetic polarization may be done by the assistance of the animal forms of which we have spoken; they will serve to fix the imagination.

Let us give an example:

You wish to act magnetically upon a person polarized like yourself, which, if you are a magnetizer, you will divine at the first contact: only that person is a little less strong than you are, a mouse, while you are a rat. Make yourself a cat, and you will capture it.

In one of the admirable stories which, though he did not invent it, he has told better than anybody, Perrault puts upon the stage a cat, which cunningly induces an ogre to change himself into a mouse, and the thing is no sooner done, than the mouse is crunched by the cat. The *Tales of Mother Goose*, like the *Golden Ass* of Apuleius, are perhaps true magical legends, and hide beneath the cloak of childish fairy tales the formidable secrets of science.

It is a matter of common knowledge that magnetizers give to pure water the properties and taste of wine, liqueurs and every conceivable drug, merely by the laying-on of hands, that is to say, by their will expressed in a sign.

One knows, too, that those who tame fierce animals conquer lions by making themselves mentally and magnetically stronger and fiercer than lions.

Jules Gérard, the intrepid hunter of the African lion, would be devoured if he were afraid. But, in order not to be afraid of a lion, one must make one's self stronger and more savage than the animal itself by an effort of imagination and of will. One must say to one's self: It is I who am the lion, and in my presence this animal is only a dog who ought to tremble before me.

Fourier imagined anti-lions; Jules Gérard has realized that chimera of the phalansterian[1] dreamer.

[1] Fourier was a Socialist who wrote a sort of 'Utopia'. His social unit was the 'phalanstère'.—A.C.

But, one will say, in order not to fear lions, it is enough to be a man of courage and well armed.

No, that is not enough. One must know one's self by heart, so to speak, to be able to calculate the leaps of the animal, divining its stratagems, avoiding its claws, foreseeing its movements, to be in a word past-master in lioncraft, as the excellent La Fontaine might have said.

Animals are the living symbols of the instincts and passions of men. If you make a man timid, you change him into a hare. If, on the contrary, you drive him to ferocity, you make a tiger of him.

The wand of Circe is the power of fascination which woman possesses; and the changing of the companions of Ulysses into hogs is not a story peculiar to that time.

But no metamorphosis may be worked without destruction. To change a hawk into a dove, one must first kill it, then cut it to pieces, so as to destroy even the least trace of its first form, and then boil it in the magic bath of Medea.

Observe how modern hierophants proceed in order to accomplish human regeneration; how, for example, in the Catholic religion, they go to work in order to change a man more or less weak and passionate into a stoical missionary of the Society of Jesus.

There is the great secret of that venerable and terrible Order, always misunderstood, often calumniated, and always sovereign.

Read attentively the book entitled, *The Exercises of St. Ignatius*, and note with what magical power that man of genius operates the realization of faith.

He orders his disciples to see, to touch, to smell, to taste invisible things. He wishes that the senses should be exalted during prayer to the point of voluntary hallucination. You are meditating upon a mystery of faith; St. Ignatius wishes, in the first place, that you should create a place, dream of it, see it, touch it. If it is hell, he gives you burning rocks to touch, he makes you swim in shadows thick as pitch, he puts liquid sulphur on your tongue, he fills your nostrils with an abominable stench, he shows you frightful tortures, and makes you hear groans superhuman in their agony; he commands your will to create all that by exercises obstinately persevered in. Every one carries this out in his own fashion, but always in the way best suited to impress him. It is not the hashish intoxication which was useful to the knavery of the Old Man of the Mountain; it is a dream without sleep, an hallucination without madness, a reasoned and willed vision, a real creation of intelligence and faith. Thenceforward, when he preaches, the Jesuit can say: 'What we have seen with our eyes, what we have heard with our ears, and what our hands have handled, that do we declare unto you.' The Jesuit thus trained is in communion with a circle of wills exercised like his own; consequently each of the fathers is as strong as the Society, and the Society is stronger than the world.

CHAPTER II

How to Preserve and Renew Youth—The Secrets of Cagliostro —The Possibility of Resurrection—Example of William Postel, Called the Resurrected—Story of a Wonder-Working Workman, etc.

ONE knows that a sober, moderately busy, and perfectly regular life usually prolongs existence; but in our opinion, that is little more than the prolongation of old age, and one has the right to ask from the science which we profess other privileges and other secrets.

To be a long time young, or even to become young again, that is what would appear desirable and precious to the majority of men. Is it possible? We shall examine the question.

The famous Count of Saint-Germain is dead, we do not doubt, but no one ever saw him grow old. He appeared always of the age of forty years, and at the time of his greatest celebrity, he pretended to be over eighty.

Ninon de l'Enclos, in her very old age, was still a young, beautiful and seductive woman. She died without having grown old.

Desbarrolles, the celebrated palmist, has been for a long while for everybody a man of thirty-five years. His birth certificate would speak very differently if he dared to show it, but no one would believe it.

Cagliostro always appeared the same age. He pretended to possess not only an elixir which gave to the old, for an instant, all the vigour of youth; but he also prided himself on being able to operate physical regeneration by means which we have detailed and analysed in our *History of Magic.*

Cagliostro and the Count of Saint-Germain attributed the preservation of their youth to the existence and use of the universal medicine, that medicament uselessly sought by so many hermetists and alchemists.

An Initiate of the sixteenth century, the good and learned William

Postel, never pretended that he possessed the great arcanum of the hermetic philosophy; and yet after having been seen old and broken, he reappeared with a bright complexion, without wrinkles, his beard and hair black, his body agile and vigorous. His enemies pretended that he rouged, and dyed his hair; for scoffers and false savants must find some sort of explanation for the phenomena which they do not understand.

The great magical means of preserving the youth of the body is to prevent the soul from growing old by preserving preciously that original freshness of sentiments and thoughts which the corrupt world calls illusions, and which we shall call the primitive mirages of eternal truth.

To believe in happiness upon earth, in friendship, in love, in a maternal Providence which counts all our steps, and will reward all our tears, is to be a perfect dupe, the corrupt world will say; it does not see that it is itself who is the dupe, believing itself strong in depriving itself of all the delights of the soul.

To believe in moral good is to possess that good: for this reason the Saviour of the world promises the Kingdom of Heaven to those who should make themselves like little children. What is childhood? It is the age of faith. The child knows nothing yet of life; and thus he radiates confident immortality. Is it possible for him to doubt the devotion, the tenderness, the friendship, and the love of Providence when he is in the arms of his mother?

Become children in heart, and you will remain young in body.

The realities of God and Nature surpass infinitely in beauty and goodness all the imagination of men. It is thus that the world-weary are people who have never known how to be happy; and those who are disillusioned prove by their dislikes that they have only drunk of muddy streams. To enjoy even the animal pleasures of life one must have the moral sense; and those who calumniate existence have certainly abused it.

High magic, as we have proved, leads man back to the laws of the purest morality. Either he finds a thing holy or makes it holy, says an adept—*Vel sanctum invenit, vel sanctum facit*; because it makes us understand that in order to be happy, even in this world, one must be holy.

To be holy! that is easy to say; but how give one's self faith when one no longer believes? How re-discover a taste for virtue in a heart faded by vice?

One must have recourse to the four words of science: to know, to dare, to will, and to keep silence.

One must still one's dislikes, study duty, and begin by practising it as though one loved it.

You are an unbeliever, and you wish to make yourself a Christian?

Perform the exercises of a Christian, pray regularly, using the Christian formulae; approach the sacraments as if you had faith, and faith will come. That is the secret of the Jesuits, contained in the Spiritual Exercises of St. Ignatius.

By similar exercises, a fool, if he willed it with perseverance, would become a wise man.[1]

By changing the habits of the soul one certainly changes those of the body; we have already said so, and we have explained the method.

What contributes above all to age us by making us ugly? Hatred and bitterness, the unfavourable judgments which we make of others, our rages of hurt vanity, and our ill-satisfied passions. A kindly and gentle philosophy would avoid all these evils.

If we close our eyes to the defects of our neighbour, and only consider his good qualities, we shall find good and benevolence everywhere. The most perverse man has a good side to him, and softens when one knows how to take him. If you had nothing in common with the vices of men, you would not even perceive them. Friendship, and the devotions which it inspires, are found even in prisons and in convict stations. The horrible Lacenaire faithfully returned any money which had been lent to him, and frequently acted with generosity and kindness. I have no doubt that in the life of crime which Cartouche and Mandrin led there were acts of virtue fit to draw tears from the eyes. There has never been any one absolutely bad or absolutely good. 'There is none good but God,' said the best of the Masters.

That quality in ourselves which we call zeal for virtue is often nothing but a masterful secret self-love, a jealousy in disguise, and a proud instinct of contradiction. 'When we see manifest disorders and scandalous sinners,' say mystical theologians, 'let us believe that God is submitting them to greater tests than those with which He tries us, that certainly, or at least very probably, we are not as good as they are, and should do much worse in their place.'

Peace! Peace! this is the supreme welfare of the soul, and it is to give us this that Christ came to the world.

[1] If the fool would but persist in his folly, he would become wise.— William Blake.

'Glory to God in the highest, peace upon earth, and good will toward men!' cried the Angels of Heaven at the birth of the Saviour.

The ancient fathers of Christianity counted an eighth deadly sin: it was Sorrow.

In fact, to the true Christian even repentance is not a sorrow; it is a consolation, a joy, and a triumph. 'I wished evil, and I wish it no more; I was dead and I am alive.' The father of the Prodigal Son has killed the fatted calf because his son has returned. What can he do? Tears and embarrassment, no doubt! but above all, joy!

There is only one sad thing in the world, and that is sin and folly. Since we are delivered, let us laugh and shout for joy, for we are saved, and all those who loved us in their lives rejoice in Heaven!

We all bear within ourselves a principle of death and a principle of immortality. Death is the beast, and the beast produces always bestial stupidity. God does not love fools, for His divine spirit is called the spirit of intelligence. Stupidity expiates itself by suffering and slavery. The stick is made for beasts.

Suffering is always a warning. So much the worse for him who does not understand it! When Nature tightens the rein, it is that we are swerving; when she plies the whip, it is that danger is imminent. Woe, then, to him who does not reflect!

When we are ripe for death, we leave life without regret, and nothing would make us take it back; but when death is premature, the soul regrets life, and a clever thaumaturgist would be able to recall it to the body. The sacred books indicate to us the proceeding which must be employed in such a case. The Prophet Elisha and the Apostle St. Paul employed it with success. The deceased must be magnetized by placing the feet on his feet, the hands on his hands, the mouth on his mouth. Then concentrate the whole will for a long time, call to itself the escaped soul, using all the loving thoughts and mental caresses of which one is capable. If the operator inspires in that soul much affection or great respect, if in the thought which he communicates magnetically to it the thaumaturgist can persuade it that life is still necessary to it, and that happy days are still in store for it below, it will certainly return, and for the man of everyday science the apparent death will have been only a lethargy.

It was after a lethargy of this kind that William Postel, recalled to life by Mother Jeanne, reappeared with a new youth, and called himself no longer anything but Postel the Resurrected, *Postellus restitutus*.

In the year 1799, there was in the Faubourg St. Antoine, at Paris, a blacksmith who gave himself out to be an adept of hermetic science. His name was Leriche, and he passed for having performed miraculous cures and even resurrections by the use of the universal medicine. A ballet girl of the Opera, who believed in him, came one day to see him, and said to him, weeping, that her lover had just died. M. Leriche went out with her to the house of death. As he entered, a person who was going out, said to him: 'It is useless for you to go upstairs, he died six hours ago.' 'Never mind,' said the blacksmith, 'since I am here I will see him.' He went upstairs, and found a corpse frozen in every part except in the hollow of the stomach, where he thought that he still felt a little heat. He had a big fire made, massaged his whole body with hot napkins, rubbed him with the universal medicine dissolved in spirit of wine. (His pretended universal medicine must have been a powder containing mercury analogous to the kermes[1] of the druggist.) Meanwhile the mistress of the dead man wept and called him back to life with the most tender words. After an hour and a half of these attentions, Leriche held a mirror before the patient's face, and found the glass slightly clouded. They redoubled their efforts, and soon obtained a still better marked sign of life. They then put him in a well-warmed bed, and a few hours afterwards he was entirely restored to life. The name of this person was Candy. He lived from that time without ever being ill. In 1845 he was still alive, and was living at Place du Chevalier du Guet, 6. He would tell the story of his resurrection to any one who would listen to him, and gave much occasion for laughter to the doctors and wiseacres of his quarter. The good man consoled himself in the vein of Galileo, and answered them: 'You may laugh as much as you like. All I know is, that the death certificate was signed and the burial licence made out; eighteen hours later they were going to bury me, and here I am.'

---

[1] Made by boiling black antimony sulphide with sodium carbonate solution. Used in gout and rheumatism and some skin diseases on the Continent, rarely in England. —A.C.

CHAPTER III

The Grand Arcanum of Death

WE OFTEN become sad in thinking that the most beautiful life must finish, and the approach of the terrible Unknown that one calls death disgusts us with all the joys of existence.

Why be born, if one must live so little? Why bring up with so much care children who must die? Such is the question of human ignorance in its most frequent and its saddest doubts.

This, too, is what the human embryo may vaguely ask itself at the approach of that birth which is about to throw it into an unknown world by stripping it of its protective envelope. Let us study the mystery of birth, and we shall have the key of the great arcanum of death!

Thrown by the laws of Nature into the womb of a woman, the incarnated spirit very slowly wakes, and creates for itself with effort organs which will later be indispensable, but which as they grow increase its discomfort in its present situation. The happiest period of the life of the embryo is that when, like a chrysalis, it spreads around it the membrane which serves it for refuge, and which swims with it in a nourishing and preserving fluid. At that time it is free, and does not suffer. It partakes of the universal life, and receives the imprint of the memories of Nature which will later determine the configuration of its body and the form of its features. That happy age may be called the childhood of the embryo.

Adolescence follows; the human form becomes distinct, and its sex is determined; a movement takes place in the maternal egg which resembles the vague reveries of that age which follow upon childhood. The placenta, which is the exterior and the real body of the foetus, feels germinating in itself something unknown, which already tends to break it and escape. The child then enters more distinctly into the life of dreams. Its brain, acting as a mirror of that of its mother, reproduces with so much force her imaginations, that it communicates their form to its own limbs. Its mother is for it at that time what God is for us, a Providence unknown and invisible, to which it aspires to the point of identifying itself with everything that she admires. It holds to her, it lives by her, although it does not see her, and would not even know how to understand her. If it was able to philosophize, it would perhaps deny the personal existence and intelligence of that mother which is for it as yet only a fatal prison and an apparatus of preservation. Little by little, however, this servitude annoys it; it twists itself, it suffers, it feels that its life is about to end. Then comes an hour of anguish and convulsion; its bonds break; it feels that it is about to fall into the gulf of the unknown. It is accomplished; it falls, it is crushed with pain, a strange cold seizes it, it breathes a last sigh which turns into a first cry; it is dead to embryonic life, it is born to human life!

During embryonic life it seemed to it that the placenta was its body, and it was in fact its special embryonic body, a body useless for another life, a body which had to be thrown off as an unclean thing at the moment of birth.

The body of our human life is like a second envelope, useless for the third life, and for that reason we throw it aside at the moment of our second birth.

Human life compared to Heavenly life is veritably an embryo. When our evil passions kill us, Nature miscarries, and we are born before our time for eternity, which exposes us to that terrible dissolution which St. John calls the second death.

According to the constant tradition of ecstatics, the abortions of human life remain swimming in the terrestrial atmosphere which they are unable to surmount, and which little by little absorbs them and drowns them. They have human form, but always lopped and imperfect; one lacks a hand, another an arm, this one is nothing but a torso, and that is a pale rolling head. They have been prevented from rising to Heaven by a wound received during human life, a moral wound which has caused a physical deformity, and through this wound, little by little, all of their existence leaks away.

Soon their moral soul will be naked, and in order to hide its shame by making itself at all costs a new veil, it will be obliged to drag itself into the outer darkness, and pass slowly through the dead sea, the slumbering waters of ancient chaos. These wounded souls are the larvae of the second formation of the embryo; they nourish their airy bodies with a vapour of shed blood, and they fear the point of the sword. Frequently they attach themselves to vicious men and live upon their lives, as the embryo lives in its mother's womb. In these circumstances, they are able to take the most horrible forms to represent the frenzied desires of those who nourish them, and it is these which appear under the figures of demons to the wretched operators of the nameless works of black magic.

These larvae fear the light, above all the light of the mind. A flash of intelligence is sufficient to destroy them as by a thunderbolt, and hurl them into that Dead Sea which one must not confuse with the sea in Palestine so-called. All that we reveal in this place belongs to the tradition of seers, and can only stand before science in the name of that exceptional philosophy, which Paracelsus called the philosophy of sagacity, *philosophia sagax.*

CHAPTER IV

Arcanum Arcanorum

THE great arcanum—that is to say, the unutterable and inexplicable secret—is the absolute knowledge of good and of evil.

'When you have eaten the fruit of this tree, you will be as the gods,' said the Serpent.

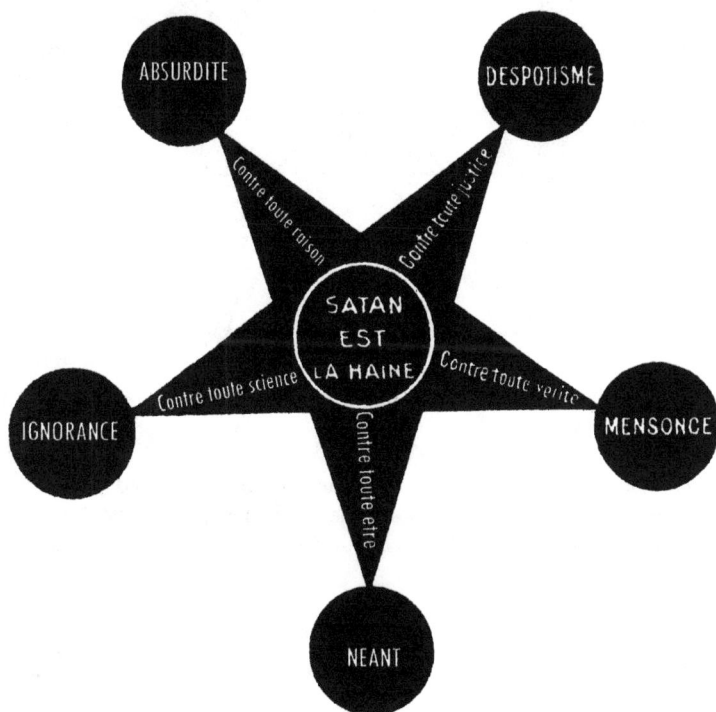

'If you eat of it, you will die,' replied Divine Wisdom.

Thus good and evil bear fruit on one same tree, and from one same root.

Good personified is God.

Evil personified is the Devil.

To know the secret or the formula of God is to be God.

To know the secret or the formula of the Devil is to be the Devil.

To wish to be at the same time God and Devil is to absorb in one's

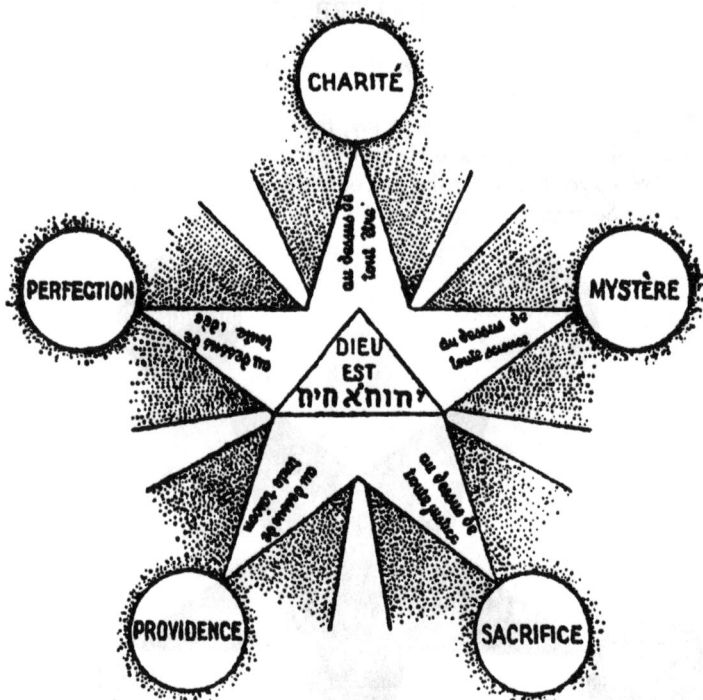

self the most absolute antinomy, the two most strained contrary forces; it is the wish to shut up in one's self an infinite antagonism.

It is to drink a poison which would extinguish the suns and consume the worlds.[1]

---

[1] An allusion to Shiva, who drank the poison generated by the churning of the 'Milk Ocean'. (See *Bhagavata Purana Skandha* VIII, Chaps. 5—12.) Lévi therefore means in this passage the exact contrary of what he pretends to mean. Otherwise this 'Be good, and you will be happy' chapter would scarcely deserve the title 'Arcanum Arcanorum' —A.C.

It is to put on the consuming robe of Deianira.

It is to devote one's self to the promptest and most terrible of all deaths.

Woe to him who wishes to know too much! For if excessive and rash knowledge does not kill him it will make him mad.

To eat the fruit of the Tree of Knowledge of Good and Evil is to associate evil with good, and assimilate the one to the other.

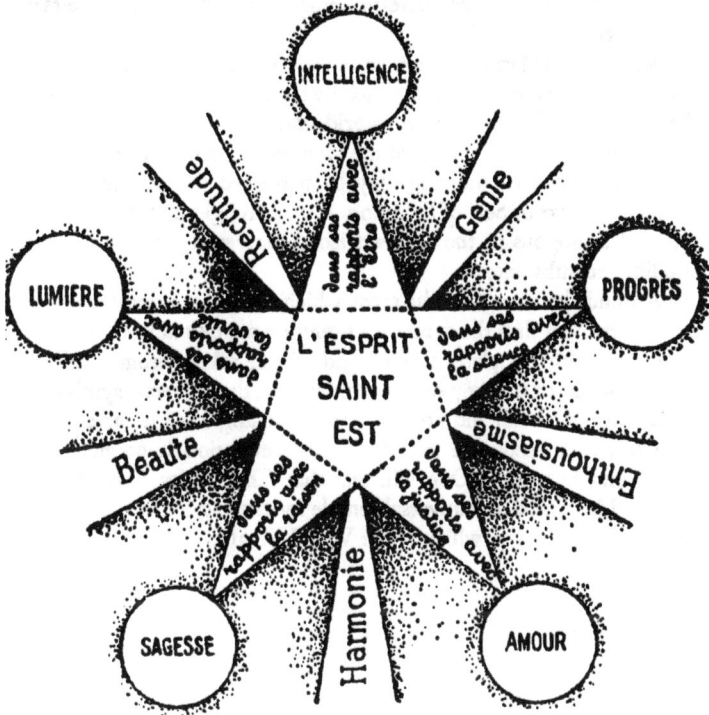

It is to cover the radiant countenance of Osiris with the mask of Typhon.

It is to raise the sacred veil of Isis; it is to profane the sanctuary.

The rash man who dares to look at the sun without protection becomes blind, and from that moment for him the sun is black.

We are forbidden to say more on this subject; we shall conclude our revelation by the figure of three pentacles.

These three stars will explain it sufficiently. They may be compared with that which we have caused to be drawn at the head of our *History of Magic*. By reuniting the four, one may arrive at the understanding of the Great Arcanum of Arcana.

It now remains for us to complete our work by giving the great Key of William Postel.

This key is that of the Tarot. There are the four suits, wands, cups, swords, coins or pentacles, corresponding to the four cardinal points of Heaven, and the four living creatures or symbolic signs and numbers and letters formed in a circle: then the seven planetary signs, with the indication of their repetition signified by the three colours, to symbolize the natural world, the human world and the divine world, whose hieroglyphic emblems compose the twenty-one trumps of our Tarot.

In the centre of the ring may be perceived the double triangle forming the Star or Seal of Solomon. It is the religious and metaphysical triad analogous to the natural triad of universal generation in the equilibrated substance.

Around the triangle is the cross which divides the circle into four equal parts, and thus the symbols of religion are united to the signs of geometry; faith completes science, and science acknowledges faith.

By the aid of this key one can understand the universal symbolism of the ancient world, and note its striking analogies with our dogmas. One will thus recognize that the divine revelation is permanent in nature and humanity. One will feel that Christianity only brought light and heat into the universal temple by causing to descend therein the Spirit of Charity, which is the Very Life of God Himself.

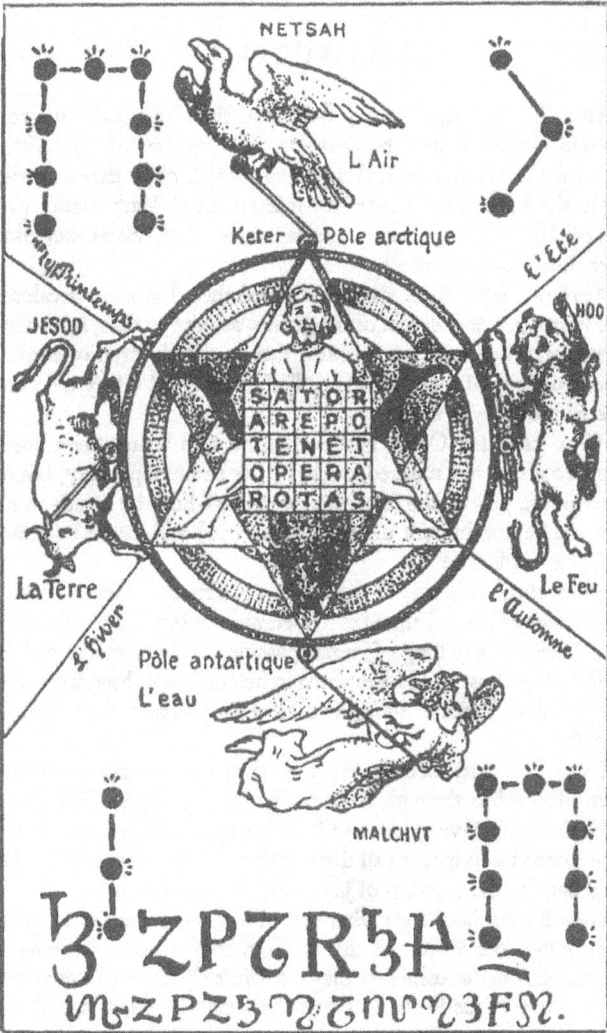

THE KEY OF WILLIAM POSTEL

EPILOGUE

THANKS be unto thee, O my God, that thou hast called me to this admirable light! Thou, the Supreme Intelligence and the Absolute Life of those numbers and those forces which obey thee in order to people the infinite with inexhaustible creation! Mathematics proves thee, the harmonies of Nature proclaim thee, all forms as they pass by salute thee and adore thee!

Abraham knew thee, Hermes divined thee, Pythagoras calculated thee, Plato, in every dream of his genius, aspired to thee; but only one initiate, only one sage has revealed thee to the children of earth, one alone could say of thee: 'I and my Father are one.' Glory then be his, since all his glory is thine!

Thou knowest, O my Father, that he who writes these lines has struggled much and suffered much; he has endured poverty, calumny, proscription, prison, the forsaking of those whom he loved—and yet never did he find himself unhappy, since truth and justice remained to him for consolation!

Thou alone art holy, O God of true hearts and upright souls, and thou knowest if ever I thought myself pure in thy sight! Like all men I have been the plaything of human passions. At last I conquered them, or rather thou hast conquered them in me; and thou hast given me for a rest the deep peace of those who have no goal and no ambition but Thyself.

I love humanity, because men, as far as they are not insensate, are never wicked but through error or through weakness. Their natural disposition is to love good, and it is through that love that thou hast given them as a support in all their trials that they must sooner or later be led back to the worship of justice by the love of truth.

Now let my books go where thy Providence shall send them! If they contain the words of thy wisdom they will be stronger than oblivion. If, on the contrary, they contain only errors, I know at least that my love of justice and of truth will survive them, and that thus immortality cannot fail to treasure the aspirations and wishes of my soul that thou didst create immortal!

INDEX